PALAEONTOLOGICAL
FIELD GUIDES TO FOSSILS: Number 6

# Plant Fossils of the British Coal Measures

*By* Christopher J. Cleal
*and* Barry A. Thomas

*Department of Botany,
National Museum of Wales,
Cardiff CF1 3NP*

THE PALAEONTOLOGICAL ASSOCIATION
LONDON
1994

© *The Palaeontological Association,* 1994

ISBN 0-901702-53-6

*Series Editor* Ed A. Jarzembowski
*P.R.I.S., The University,
Reading RG6 2AB*

Printed in Great Britain by Henry Ling Ltd., at the Dorset Press, Dorchester, Dorset

# CONTENTS

| | |
|---|---|
| INTRODUCTION | 5 |
| STRATIGRAPHICAL BACKGROUND | 6 |
| PALAEOECOLOGY OF THE COAL MEASURES | 12 |
| PLANT RECONSTRUCTIONS | 21 |
| TAPHONOMY | 37 |
| COLLECTION AND CONSERVATION OF SPECIMENS | 43 |
| FOSSIL LOCALITIES | 44 |
| INTRODUCTION TO THE KEYS | 46 |
| KEY TO GROUP A: STEMS OR STEM-LIKE FOSSILS WITH LONGITUDINAL STRIATIONS OR RIBS | 48 |
| KEY TO GROUP B: AXES WITH SCARS DISTRIBUTED REGULARLY IN VERTICAL ROWS OR SPIRALS | 57 |
| KEY TO GROUP C: LEAVES BORNE ON STEMS IN WHORLS | 89 |
| KEY TO GROUP D: BRANCHING STEMS COVERED WITH SMALL, TAPERING OR STRAP-LIKE 'LEAVES', BELONGING TO THE LYCOPSIDS AND CONIFERS | 98 |
| KEY TO GROUP E: PARTS OF FERN-LIKE FRONDS | 104 |
| KEY TO GROUP F: CIGAR-SHAPED 'CONES' OR STROBILI BORNE SINGLE OR IN A SERIES ALONG A STEM | 159 |
| KEY TO GROUP G: DETACHED LYCOPSID CONE-SCALES OR LEAVES | 167 |
| KEY TO GROUP H: SEEDS AND BELL-SHAPED CLUSTERS OF POLLEN SACS | 171 |
| BIOSTRATIGRAPHY | 182 |
| GLOSSARY | 194 |
| REFERENCES | 198 |
| TAXONOMIC INDEX | 211 |

# ACKNOWLEDGEMENTS

Dr. D. M. Spillards prepared the text-figures and the plates for this Field Guide. Without her assistance this project would never have been completed. We also thank the Keeper of Geology (National Museum of Wales), the Keeper of Palaeontology (Natural History Museum, London), the Director (British Geological Survey) and the Assistant Curator of Natural History (Tolson Memorial Museum, Huddersfield) for giving us access to specimens in their collections, and the Photography Department (National Museum of Wales) for photographic assistance. Last, but not least, thanks to all the collectors who supplied us with material over the years.

# INTRODUCTION

The Carboniferous Coal Measures have yielded some of the most abundant and diverse plant fossil assemblages in Britain. About 300 species have been recorded, providing an important insight into the nature of the palaeoequatorial vegetation of 300 million years ago. They have been investigated since the seventeenth century (for a review of the historical background, see Andrews 1980), but remain surprisingly poorly understood. Other than the well illustrated studies by Kidston (1923–1925) and Crookall (1955–1976), there are few adequately documented records; even the classic Radstock and Barnsley Seam assemblages have never been properly monographed. Thus, much work still needs to be done in documenting these plant fossils, and there is a particular urgency because of the progressive closure of many of the collieries and the landscaping of their spoil tips, which are the best sources of these fossils.

The fossils can be studied at various levels. They can be investigated just for their own sake—sometimes rather disparagingly referred to as 'stamp collecting' (e.g. Boulter *et al.* 1991), but in fact a perfectly legitimate branch of natural history, on a par with botany or mineralogy. They can also be seen as geological tools, having proved to be of great value for establishing detailed stratigraphical correlations, and helping understand the configuration of the continents at that time (Wagner 1984; Cleal 1991*a*; Cleal and Thomas 1991). From an evolutionary perspective, they are of interest as representing mostly more primitive groups of plants than are found in comparable habitats today. Although mostly trees, many had quite different growth and reproductive strategies from modern trees. They represent part of the first extensive tropical forest ecosystem to develop on earth; the growth and eventual extinction of this forest, and its possible impact on the climatic changes that occurred towards the end of the Carboniferous, has a clear relevance to the current debate on the impact of Recent tropical forest destruction.

Some of this work (e.g. cuticle studies) requires specialist equipment, and is beyond those without access to laboratory facilities. There remains, however, much that can be done with little more than a hammer and a hand-lens. This guide has been written to assist those who wish to investigate Coal Measures plant fossils, particularly in helping with their identification. It is only intended as a guide to their preliminary identification, and anyone wanting to make a detailed investigation will have to delve further into the scientific literature. It is hoped, however, that it will be helpful to those unfamiliar with these fossils, and provide an introduction to this interesting branch of palaeontology.

# STRATIGRAPHICAL BACKGROUND

The British Coal Measures belong to the Westphalian and basal Stephanian series of the Carboniferous. The group occurs extensively in the British Isles, with an area of outcrop of well over 7000 km$^2$, and an even larger area of subsurface strata. They include twenty-one major coalfields, whose distribution is mainly constrained by three prominent geological structures: the Variscan Front, St. George's Land (or the Wales-Brabant Massif), and the Highland Boundary Fault. The major coalfields are shown in Text-figure 1.

Following current stratigraphical practice (Hedberg 1976) these strata can be separately classified by time (chronostratigraphy), fossil content (biostratigraphy) and litho-facies (lithostratigraphy).

*Chronostratigraphy*
During the Late Carboniferous, the Earth developed distinct climatic belts (Rowley *et al.* 1985), which in turn resulted in marked faunal and floral provincialism (Cleal and Thomas 1991). This has made it difficult to establish a global chronostratigraphical classification for strata of this age, and so different schemes have been developed for use in different parts of the world. There are moves afoot to establish more universally applicable schemes for the Upper Carboniferous (Bouroz *et al.* 1978; Winkler Prins 1989; Brenckle and Manger 1991) but such proposals are still very much in the planning stage and cannot yet be regarded as practical stratigraphical tools.

For much of Europe and eastern Canada, the 'Heerlen Classification' is generally used. It is named after the early conferences on Carboniferous stratigraphy held at Heerlen, in the Netherlands (Jongmans 1928;

---

TEXT-FIG. 1. Map of principal outcrops of Coal Measures in the British Isles. 1, Inninmore; 2, Machrihanish; 3, Arran; 4, Ayrshire Coalfield; 5, Sanquhar Coalfield; 6, Douglas Coalfield; 7, Central Coalfield; 8, Lothians Coalfield; 9, Canonbie; 10, West Cumberland Coalfield; 11, Midgeholm; 12, Stainmore; 13, Northumberland-Durham Coalfield; 14, Ingleton; 15, Lancashire Coalfield; 16, Yorkshire and East Midlands Coalfields; 17, Flint Coalfield; 18, Denbigh Coalfield; 19, Shrewsbury Coalfield; 20, Coalbrookdale Coalfield; 21, Wyre Forest; 22, South Staffordshire Coalfield; 23, North Staffordshire Coalfield; 24, Leicestershire–South Derbyshire Coalfield; 25, Warwickshire Coalfield; 26, Pembrokeshire; 27, South Wales Coalfield; 28, Forest of Dean; 29, Bristol-Somerset Coalfield (including the concealed portion, shown stippled); 30, Coalisland; 31, Kingscourt; 32, Leinster; 33, Slieveardagh; 34, Crataloe; 35, Kanturk; 36, Bideford; 37, Anglesey.

# Stratigraphical Background

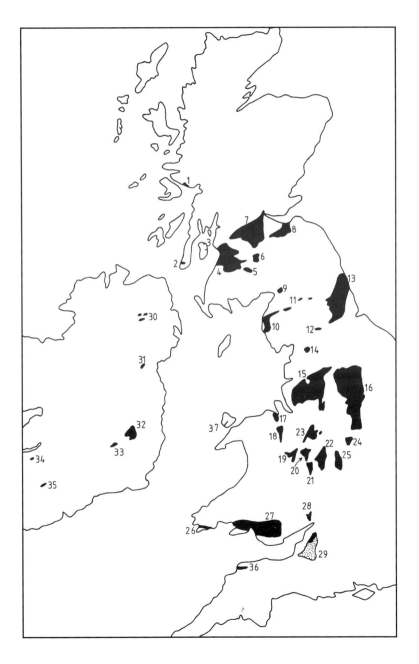

| Old Scheme | | New Scheme | |
|---|---|---|---|
| Stages | Substages | Series | Stages |
| Stephanian (*pars*) | Stephanian A | Stephanian (*pars*) | Barruelian |
| | | | Cantabrian |
| Westphalian | Westphalian D | Westphalian | 'Westphalian D' |
| | Westphalian C | | Bolsovian |
| | Westphalian B | | Duckmantian |
| | Westphalian A | | Langsettian |
| Namurian (*pars*) | Namurian C | Namurian (*pars*) | Yeadonian |

TEXT-FIG. 2. The chronostratigraphy of the British Coal Measures, showing both the old and the new versions of the Heerlen Classification.

Jongmans and Gothan 1937; van der Heide 1952; van Leckwijck 1960) where the outlines of the scheme were established. Some minor changes to it have since been made (Wagner 1974; Owens *et al.* 1985; Wagner and Winkler Prins 1985), but its essential structure remains intact. There are two subsystems: the Dinantian (the lower) and the Silesian (the upper). Only the latter is relevant to this guide, whose subdivisions are shown in Text-figure 2. Ramsbottom *et al.* (1978) proposed to subdivide the Westphalian stages into chronozones (based on the non-marine bivalve biozones) but this has doubtful relevance to the present work.

The names and boundary stratotypes for the lower three stages of the Westphalian have now been formally ratified by the IUGS Subcommission on Carboniferous Stratigraphy (Manger 1985; Engel 1989). The stage boundaries are placed at thin, discrete marine bands, which have been identified over large areas of western and central Europe: the Subcrenatum, Vanderbeckei and Aegiranum marine bands (Owens *et al.* 1985). No such marine horizons occur above the middle Bolsovian and so the higher stage boundaries are recognized on biostratigraphical criteria, principally palaeobotanical. There is as yet no boundary stratotype for the Westphalian D, although South Wales (Cleal 1978) and northwest Spain (Wagner and Alvarez-Vázquez 1991) are possible contenders. Nevertheless, it is generally recognized that the base of the *Linopteris obliqua* Zone can be used as an index for the base of the Westphalian D (Cleal 1984*a*). The stratotypes for the Cantabrian and Barruelian stages are both in northwest Spain (Wagner 1984; Wagner *et al.* 1983; Wagner and Winkler Prins 1985) and their bases can be identified by the base of the *Odontopteris cantabrica* and *Lobatopteris lamuriana* zones respectively.

Stratigraphical Background

| STAGES | PLANT MACROFOSSILS ZONES | PLANT MACROFOSSILS SUBZONES | NON-MARINE BIVALVES ZONES | NON-MARINE BIVALVES SUBZONES | PALYNOLOGY ZONES |
|---|---|---|---|---|---|
| Cantabrian | *Odontopteris cantabrica* | | | | |
| 'Westphalian D' | *Lobatopteris* | *Dicksonites plueckenetii* | *Anthraconauta tenuis* | | *Thymospora obscura* |
| | | *Lobatopteris micromiltoni* | | | |
| | *Linopteris bunburii* | | | | |
| Bolsovian | *Paripteris linguaefolia* | *Alethopteris serlii* | *Anthraconauta phillipsi* | | *Torispora securis* |
| | | *Lavineopteris rarinervis* | | | |
| | | *Neuropteris semireticulata* | 'Upper similis-pulchra' | *A. adamsi - A. hindi* | |
| Duckmantian | *Lonchopteris rugosa* | *Sphenophyllum majus* | 'Lower similis-pulchra' | *Anthracosia atra* | *Microreticulatisporites nobilis - Florinites junior* |
| | | *Neuropteris hollandica* | *Anthraconaia modiolaris* | *Anthraconaia caledonica* | |
| | | | | *Anthracosia phrygiana* | |
| | | | | *Anthracosia ovum* | |
| Langsettian | *Lyginopteris hoeninghausii* | *Lavineopteris loshii* | *Carbonicola communis* | *Anthracosia regularis* | *Radizonates aligerens* |
| | | | | *Carbonicola cristagalli* | |
| | | | | *Carbonicola pseudorobusta* | |
| | | | | *Carbonicola bipennis* | |
| | | *Neuralethopteris jongmansii* | | *Carbonicola torus* | |
| | | | *Carbonicola lenisulcata* | *Carbonicola proxima* | *Triquitrites sinanii - Cirratriradites saturnii* |
| | | | | *Carbonicola extenuata* | |
| | | | | *C. fallax - C. proiea* | |

TEXT-FIG. 3. The biostratigraphy of the British Coal Measures, showing relative positions of the zones for plant macrofossils, non-marine bivalves and palynology.

*Biostratigraphy*
Three main groups of fossils have been used for biostratigraphy in the Coal Measures: non-marine bivalves, plant macrofossils, and pollen and spores. In the Westphalian and Lower Stephanian, seven non-marine bivalve zones, six pollen and spore zones and seven plant macrofossil zones (plus nine subzones) have been recognized (Text-fig. 3 shows the Langsettian to Cantabrian zones). Further details of the plant macrofossil biozonation are supplied below (see also Cleal 1991*a*). Certain other fossil groups have been investigated as potential biostratigraphical tools, including vertebrates (Carroll 1984) and insects (Schneider 1983; Durden 1984), but as yet none has provided as fine a resolution as the above groups.

*Lithostratigraphy*
Most Westphalian and Lower Stephanian strata in Britain are traditionally called the Coal Measures. There have been few attempts to categorize the Coal Measures in terms of the lithostratigraphical classificatory hierarchy proposed by Hedberg (1976), but it seems to fit most closely with the concept of a group. The base of the Coal Measures is now placed at the base of the Subcrenatum Marine Band (Stubblefield and Trotter 1957) and thus coincides with the base of the Westphalian Series.

Several subdivisions of the Coal Measures Group have been recognized, reflecting the facies variations that occur across the country (for further details of the facies variations in the British Coal Measures, see Ramsbottom *et al.* 1978). What are often thought of as the typical Coal Measures—the mainly grey, coal-bearing mudstones and shales, with usually thin sandstones—are often referred to as the Productive Coal Formation. Such strata are widely developed in the Langsettian to middle Bolsovian in most of the British coalfields and provide the best source of plant fossils dealt with in this volume.

However, the effects of the Variscan earth movements resulted in widespread facies changes, mainly in the middle and upper Westphalian. Uplift of the depositional areas resulted in falling water tables and the development of red beds (Besly 1988). These are typically developed in the English Midlands, where the diachronous Etruria Formation ranges from the Langsettian to lower Westphalian D. Other well-documented developments of red beds are in the Bolsovian and Westphalian D of the Scottish coalfields, and in the upper Westphalian D of the Somerset Coalfield, both of which are termed the Barren Red Formation. Such red beds do not often yield abundant plant fossils, although where they do occur, they can be of interest as representing the vegetation of somewhat drier habitats, including the remains of plants such as conifers (Lyons and Darrah 1989).

Another response to these Variscan movements was uplift of the hinterland areas, causing increased erosion and the introduction of more arenaceous sediments. These are typically developed in South Wales, the Forest of Dean and the Bristol-Somerset coalfields, where they are known as the Pennant Formation (Kelling 1974; Cleal 1991*b*). Less well documented are major sandstones in the English Midlands, known as the Halesowen Formation, and in the West Cumberland Coalfield, known as the Whitehaven Sandstone. In the Midlands, there are also red sandstones known as the Keele Beds (in fact, a series of quite discrete red sandstone and mudstone deposits—B. Besly, pers. comm.). Such sandstones themselves rarely yield good plant fossils, but shale intercalations within the sandstones can produce good material (e.g. Cleal 1978).

In some areas, such as the West Cumberland Coalfield, these sandstones mark the end of Carboniferous deposition. Especially in southern Britain, however, there is often a return to coal-bearing deposits in the top Westphalian D and Cantabrian. Examples include the Grovesend Formation in South Wales, the Suprapennant Formation in the Forest of Dean, and the Farrington and Radstock formations in the Bristol-Somerset Coalfields. The latter are well known for producing finely-preserved plant fossils.

# PALAEOECOLOGY OF THE COAL MEASURES

*Physical environment*
During the Late Carboniferous, much of Britain was part of the foreland (used here in its traditional sense—e.g. Krausse and Pilger 1977) of the Variscan orogenic belt that resulted from the collision between the Gondwana and Laurasia continental masses. It lay only a few degrees north of the equator and was subject to a wet-tropical, possibly monsoonal climate (Rowley *et al.* 1985; Broadhurst 1988).

The British Coal Measures represent mainly alluvial sediments deposited on this foreland area (Guion *in* Cope *et al.* 1992). They formed part of an elongate, forest-covered delta-plain that extended between Poland and the British Isles, and which is sometimes referred to as the European Paralic Coal Basin. The sediment was derived from several sources. In areas between southern Scotland and the northern English Midlands, the sediment was deposited from rivers flowing from the Caledonian Uplands to the north (Guion and Fielding 1988). The Wales-Brabant Barrier was a sediment-source for South Wales and the Bristol-Somerset area to the south, and to a lesser extent for the English Midlands to the north. However, it was never a major source and probably ceased to be a topographic high altogether towards the end of the Westphalian. A third source of sediment was an upland area in what is now the Bristol Channel, and which has been called Sabrina (Kelling 1974). Today, the area does not look particularly large. However, it is thought to have been subject to considerable crustal shortening and, in view of the quantity of sediment that was derived from there (e.g. the Pennant Formation of the South Wales and Bristol-Somerset coalfields), must at the time have been a far more extensive area.

Current views on the sedimentology of the Coal Measures came initially from facies analyses by Elliott (1968, 1969), and have been further developed by, among others, Scott (1978, 1979), Haszeldine (1983*a*, 1983*b*, 1984), Fielding (1984*a*, 1984*b*, 1986, 1987), Guion (1984, 1987*a*, 1987*b*) and Guion and Fielding (1988). In essence, delta-plains were being dissected by a complex set of distributary channels, which varied considerably in their degree of sinuosity, but which were normally constrained by high banks (or levees). The levee-bounded areas between the channels, known as floodbasins, were for most of the time covered by swamp forest. The bulk of the sediments are non-marine, alluvial deposits, but there are also at least nineteen thin marine bands. They reflect periodic flooding of the delta-plain, probably as a result of melting of the polar ice-

## Palaeoecology of the Coal Measures

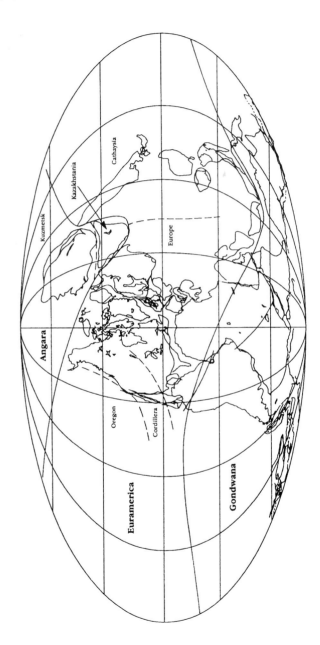

Latest Carboniferous (Stephanian)

TEXT-FIG. 4. World palaeogeography of the Late Carboniferous, showing distribution of main palaeofloristic zones. From Cleal and Thomas (1991).

caps raising sea-levels (Ramsbottom 1979; Leeder 1988). They confirm that the European Paralic Coal Basin was open to the sea, in contrast to areas such as the Saar-Lorraine and Central Bohemian basins in central Europe, which appear to have been entirely enclosed, limnic basins. These flood-events were mainly in the early Langsettian, and in the late Duckmantian to early Bolsovian, during which times the British Coal Measures may be seen as lower delta-plain in character (Fielding 1987). Other parts of the Coal Measures, especially the Westphalian D, are more upper delta-plain in character (Fielding 1984a; Besly 1988).

TEXT-FIG. 5. Reconstructions of part of the Coal Measures swamps. Above, when floodbasins were mostly covered by peat-forming forests. Below, during times of flood. From Thomas and Cleal (1993a) and taken from animated film loop prepared for the Evolution of Wales exhibit, National Museum of Wales, Cardiff.

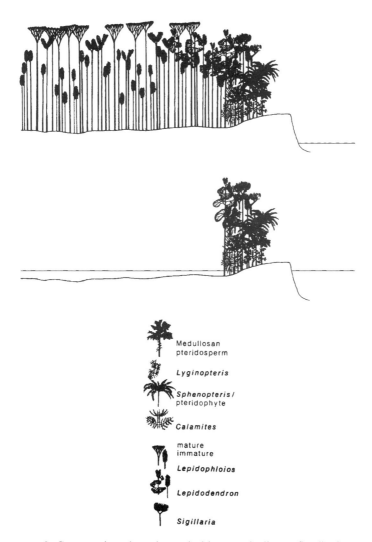

TEXT-FIG. 6. Cross sections through a typical levee and adjacent floodbasin area of a Coal Measure swamp, (upper) when floodbasins were mostly covered by peat-forming forests, and (lower) during times of flood. Adapted from Gastaldo (1987).

The trees of the floodbasin swamps produced considerable quantities of plant litter. There was also some clastic sediment introduced into the floodbasins, probably as a result of periodic overbank flooding. Normally, the build-up of plant litter outstripped the deposition of sediment, with the result that peat deposits were generated, that eventually changed into coal. These swamps with a peaty substrate are known as coal swamps. Sometimes, however, more sediment was deposited than plant litter, preventing the build-up of peat, and resulted in clastic swamps (Gastaldo et al. 1989).

The river water and some of its sediment would eventually flow into the sea, possibly somewhere to the southwest, but at least some of the sediment was deposited within the delta-plain. At times of high discharge, the coarser sediment would be deposited within the distributary channels as various types of bar, e.g. the upper Westphalian sandstones of South Wales known as the Pennant Formation. More typically, however, deposition occurred in the floodbasins. At times of flood, breaches or crevasses would develop in the levees, causing a lake to form in the floodbasin and destroying the forest growing there (Text-figs 5–6). As the moving water from the distributary channel met the standing water of the lake, it caused the sediment to settle out, and thus a small delta to grow out from the crevasse. Immediately adjacent to the crevasse, in what is known as the crevasse channel, the sediment would often be coarse but further away from the crevasse, in the crevasse splay, the sediment was predominantly fine—producing the grey shales and mudstones typical of the Coal Measures. Eventually, this crevasse splay sediment would completely silt up the floodbasin lake, allowing the forest to return. This cyclicity of flooding of the floodbasin, the silting-up of the resulting lake and the return of the forest, is the essence of much of the Coal Measures depositional pattern, and used to be referred to as 'cyclothems' (e.g. Woodland and Evans 1964).

A typical Coal Measures sequence consists predominantly of clastic rocks, with only subsidiary coals, but this provides a very misleading impression of the relative times represented by the two types of deposit. It has been estimated that the peat needed to generate a metre thickness of coal could have taken anything up to 7000 years to be laid down, whereas the same thickness of clastic sediments may have been deposited in as little as five years (Broadhurst et al. 1980; Broadhurst and France 1986). Using these figures, it becomes clear that for well over 95% of the time the delta-plain was covered by forest, and that the lakes, where much of the clastic sediment was laid down, were relatively short-lived.

*Vegetational habitats*
There has been a long history of trying to correlate plant fossil distribution and Upper Carboniferous sedimentological facies (reviewed by Scott

1977), but trying to trace this information back through uncertain transportation histories to determine the structure of original plant communities is fraught with hazards (Scott 1979). Real progress did not take place until the distribution of *in situ* stumps was investigated (Gastaldo 1985, 1987). This showed that the fragments of mainly foliage and fructifications found in the floodbasin sediments (i.e. the type of fossils dealt with in the present volume) bear little relationship to the original vegetation of the floodbasins, but was rather derived from the vegetation growing on the levees of the distributary channels. This is further supported by comparisons with Recent analogues of the Late Carboniferous delta-plain deposits, such as in Alabama (Gastaldo *et al.* 1987) and the Orinoco of Venezuela (Scheihing and Pfefferkorn 1984). A useful review of the general problems inherent in ecological interpretation of plant fossil distribution is provided by Spicer (1989).

In the currently accepted model, two main vegetational habitats can be recognized in the delta-plain forests: the floodbasins and the levees. When a floodbasin was flooded, thickets of calamite horsetails grew around the margins of the lakes. As sediment filled up the lakes, however, the area was invaded by dense swamp-forests. For much of the Westphalian, the floodbasin swamp-forests of Britain were probably dominated by lycopsid trees. Recent work on the distribution of these lycopsids within the swamp, based on petrifactions from the coal seams (DiMichele and Phillips 1985; Phillips and DiMichele 1992), suggests that a complex ecosystem was present. The floodbasins were first occupied by pioneers such as *Paralycopodites*, which were adapted to open areas with nutrient-rich, clastic substrates. As the swamp became more established, they were replaced by other lycopsids such as *Diaphorodendron*, which were better suited to the nutrient-poor, peaty substrates, and more stable setting of the more mature forests. There were also opportunistic species which could occupy areas that suffered from brief periods of flooding, such as *Synchysidendron* and *Lepidodendron* that favoured the clastic swamps, and *Lepidophloios* that favoured the peat-accumulating swamps. Evidence from North America suggests that cordaites may have occupied drier parts of the swamp (Phillips and Peppers 1984; Phillips *et al.* 1985) but direct evidence of this from Europe is lacking.

In the early Stephanian, there was a significant shift in the balance of the swamp vegetation, with tree-ferns taking over from the lycopsids as the dominant component (Phillips and Peppers 1984; Phillips *et al.* 1985). This is thought to reflect a change to the somewhat drier climatic conditions that characterize much of the Stephanian. There has not been the same detailed palaeoecological work done on these later coals, and so the ecological dynamics of the later forests is not as well understood.

The levees, because of their higher elevation, favoured a different type of vegetation. There was not a sharp demarcation between the floodbasin and levee habitats, an intermediate stage dominated by somewhat smaller lycopsids such as *Sigillaria* having been recognized (Gastaldo 1987). However, the levee habitat proper contained a much more diverse assemblage than the floodbasin swamps, including ferns, pteridosperms, cordaites and sphenopsids. It included trees of varying size, shrubs, herbaceous forms, ground-creepers and lianas. However, the exact distribution of these different plants on the levees is still unknown. Their fossilized remains are usually fragments transported by winds and floodwaters into the muds and sands of the floodbasin lake, making it difficult to relate them to their original position of growth. It is reasonable to assume, however, that there was a complex of pioneer, 'site occupier' and opportunistic sub-assemblages, such as seen in the floodbasin swamp-forests.

Although only a small part of the overall plant biosystem of the deltas, the levee assemblage usually predominates in the adpression fossil record of the Coal Measures. The greater diversity of this assemblage resulted in higher interspecific competition than in the floodbasin forests, with a consequential higher rate of species evolution. This has made the fossils originating from the levee vegetation far more useful for biostratigraphical work (Wagner 1984; Cleal 1991*a*) than those of the swamp-forest, in which the major changes in composition were triggered mainly by external climatic influences (Phillips and Peppers 1984; Phillips *et al.* 1985; DiMichele *et al.* 1985).

The drier, so-called 'upland' or extra-basinal areas surrounding the deltas provided yet another habitat for plants. The nature of the vegetation occupying this habitat has been the subject of some debate (Peppers and Pfefferkorn 1970; Havlena 1971; Knight 1974; Leary 1975, 1977; Lyons and Darrah 1989) but, until recently, the only evidence from Britain was from pollen and spores (Chaloner 1958), whose affinities were not always clear. Charred fragments macerated from Coal Measures sediments suggest that conifers may have occupied this higher ground (Scott and Chaloner 1983).

The effects on the forests of the flooding by eustatic sea-level rises seems to have been variable. The depth of the water produced by these floods varied considerably (Calver 1968), but even a shallow inundation by saline water would have been expected to have had a major impact on the plant life of the forests. Species which reproduced by spores would presumably have been able to survive in the distant intra-montane basins (e.g. the Saar-Lorraine Basin), and could possibly then have recolonized the deltas after the retreat of the flood-water. The impact on the gymnosperms, however, should have been much greater, as there is little evidence that animals provided any long-distance transportation of the seeds. The Subcrenatum Marine Band indeed seems to mark a significant change in the composition of the vegetation (Wagner 1984), but the others

Palaeoecology of the Coal Measures   19

appear to have had little or no impact. This may indicate that the viability of the seeds was not seriously damaged by immersion in saline water; or that the marginal areas of the deltas which were not flooded retained remnant populations of plants that recolonized the deltas after the retreat of the sea-water.

*Animal life*

This is not the place to go into details about the animal life of the Late Carboniferous tropical forests. On the other hand, even at this relatively early stage in the evolution of terrestrial biotas, there was a close interaction between animal and plant life (Scott 1980; Scott and Taylor 1983; Taylor and Scott 1983), and so the former cannot be totally ignored in a discussion on the vegetation. Unfortunately, there has been no synthesis of Coal Measures palaeozoology published in recent years, to which the reader could be directed for further details.

The aqueous habitats contained various faunas, containing fish such as freshwater sharks, arthropods such as horseshoe crabs and microscopic Crustacea, and bivalves. This fauna appears to have had little direct interaction with the living vegetation, but plant debris may have formed the base of some of the food chains (Scott 1980; Milner 1980). Also, the prevalence of peaty substrates may well have influenced the chemistry of the waters in which the animals lived.

Vertebrates spending most or all of their life cycle on land at this time were either amphibians or reptiles (Panchen 1970; Milner and Panchen 1973; Milner 1980; Cox *et al.* 1988). They appear to have been nearly all carnivores, although vegetation clearly had some influence on their lifestyle by generating suitable habitats and, in some cases, providing shelter and protection from predation.

Terrestrial invertebrates appear to have been mainly arthropods (Rolfe 1980, 1986). This is excluding those without hard parts, such as nematodes, for which we have little knowledge other than from trace fossils; also land snails, whose presence in the Coal Measures has not been conclusively proved. By far the most numerous arthropods, at least as judged by their representation in the fossil record, were insects (Jarzembowski 1987); nearly two-thirds of the animal fossils found at a particularly rich faunal site in Avon were reported to be insects (Jarzembowski 1989). Although many of these insects belong to extinct groups, some may be referred to as cockroaches and dragonflies. It is claimed that many were herbivorous, adapted to consuming sap or pollen/spores (Smart and Hughes 1973). There are also several examples of insect coprolite reported containing leaf and other plant remains (e.g. Scott and Taylor 1983), but there is little conclusive evidence of foliage damage that can be ascribed to insects, despite many of the leaves having been only thinly cutinized (Cleal and Shute 1991). Those reported examples of 'chewed' leaves may well represent damage inflicted after they had become leaf litter. Certainly many, especially of the cockroach-types, probably lived among the vegetation litter; it has been argued by Taylor and Scott (1983) that the similarity of their wings to certain pteridosperm pinnules was a camouflage against predators, although Jarzembowski (1987) has queried this idea.

Various groups of arachnids are known, including spiders, scorpions, whip scorpions, solpugids and ricinuleids (Petrunkevitch 1953). They were almost certainly all carnivores, but would have used vegetation in their hunting strategies—spiders spinning simple traps in the larger, compound leaves, and the ricinuleids 'skulking motionless under rotten logs, behind leaf fronds on wet mud and in wet leaf-axils . . . emerging only to catch and eat living spider and insect prey' (Rolfe 1980).

Other arthropods living in the Late Carboniferous tropical forests included mites, springtails and millipedes. Living representatives of all these groups include herbivores, although the diet of the Carboniferous forms is mostly not known. One notable exception is the giant (?)millipede *Arthropleura*, from whose gut Scott *in* Rolfe (1980) reported tracheids of arborescent lycopsids.

# PLANT RECONSTRUCTIONS

In the pages which follow, there are reconstructions of some of the major groups of plants which flourished in the British Coal Measures forests. They do not include every type of plant growing in these forests, but should be regarded as illustrative examples, which attempt to bring the vegetation alive for the reader. The accompanying comments are not an exhaustive account of these plants, but will provide a general background to their botanical affinities and to the fossils which they produced. Further details may be found in Thomas and Spicer (1987).

## Lycopsids

Some of these plants are the largest that lived in the Late Carboniferous tropical forests. They formed extensive forests in the lowland swamps; their stems and reproductive organs are common fossils in the roof-shales and sandstones, and their rooting organs are found in the seat-earths. They are related to the Recent club-mosses and quilworts but, in contrast to the modern herbaceous forms, many were large trees. An example of a mature, reconstructed lycopsid is shown in Text-figure 7 and others can be found in Phillips and DiMichele (1992). The largest grew up to 50 m tall, supporting themselves by extensive secondary cortex rather than by wood formation. Stems therefore show signs of cortical and epidermal expansion, and of a progressive loss of the outermost tissues of the stem (Text-fig. 8). The commonest form-genus of stems is *Lepidodendron*, which has diamond-shaped, swollen basal portions of the leaves (leaf cushions), upon which can often be seen a scar marking the area of attachment of the distal leaf lamina (Text-fig. 9). Leaves are rarely found attached to the larger stems, but this may be just a function of preservation (Leary and Thomas 1989). Fossils representing terminal shoots usually still bear the leaves and sometimes have terminally attached cones (Text-fig. 10) belonging to the form-genera *Flemingites* (bearing both microspores and megaspores), *Lepidostrobus* (bearing only microspores) and *Lepidocarpon* or *Achlamydocarpon*. The latter two, known mainly from petrifactions, have only a single megaspore in each sporangium. Isolated leaves are called *Cyperites* and sporophylls *Lepidostrobophyllum*. The basal rooting organs (or rhizophores) known as *Stigmaria* were shallow structures, particularly well suited to a plant growing in swamp conditions. Further details about this group can be found in Thomas (1978, 1981) and Phillips and DiMichele (1992).

The growth of these large lycopsids was significantly different from modern trees, and it is arguable that they should not really be called trees

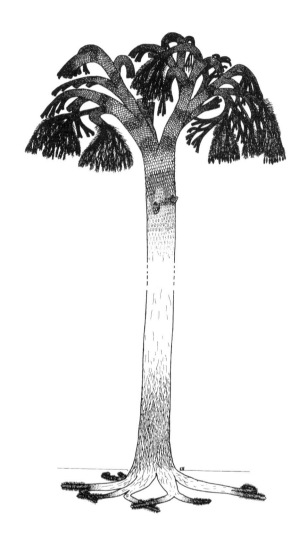

TEXT-FIG. 7. Reconstruction of the arborescent lycopsid *Lepidodendron*, about 40 m tall. From Thomas and Spicer (1987).

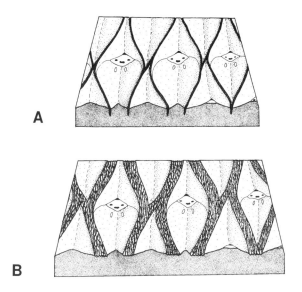

TEXT-FIG. 8. *Lepidodendron* leaf cushions before (A) and after (B) the expansion of the stem's tissues following secondary growth. After Thomas (1966).

at all (Phillips and DiMichele 1992). The rooting structure produced a vertical leafy trunk or 'pole' which had no branches. This trunk grew until just before maturity, and only then was a tree-like crown of branches produced, such as shown in Text-figure 7. The economic construction of these 'poles', with their low proportion of secondary wood, meant that they could grow very quickly; Phillips and DiMichele (1992) estimated that their life span would probably be no more than 10–15 years. Also, because they did not normally have a crown or branches except at the very end of their life, the forests were probably quite light areas.

Another stem form-genus with prominent leaf scars is *Lepidophloios*, which had bulging and drooping leaf cushions on all but its terminal shoots. In *Bothrodendron*, the leaf scars are usually flush with the stem surface, with leaf cushions only occurring on its smallest shoots. *Sigillaria* had its leaf scars secondarily arranged in vertical rows, and bore cones known as *Sigillariostrobus* and leaves called *Sigillariophyllum*.

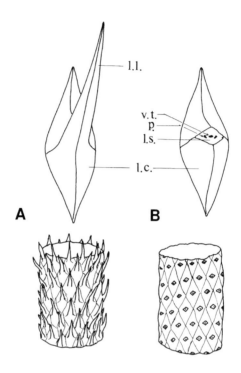

TEXT-FIG. 9. Leaf abscission in *Lepidodendron*. A, leaf cushion with leaf lamina still attached. B, the same after leaf abscission. From Thomas and Spicer (1987). Abbreviations: l.c., leaf cushion; l.l., leaf lamina; l.s., leaf scar; v.t., vascular trace; p., parichnos.

## *Calamites*

These were very similar to living horsetails (*Equisetum*), except that they were rather larger (Text-fig. 11), and formed dense thickets along the edges of waterways and lakes. As in the modern forms, the vast majority of them had extensive lateral rhizomes, which extended over the mud or were shallowly buried, and which bore the vertically growing stems. The stems (*Calamites*) branched regularly and the various branching patterns have been used to distinguish sub-form-genera. The whorled leaves on the smaller branches are called *Asterophyllites* or *Annularia*. Compression fossils of the cones are assigned to *Calamostachys*, *Palaeostachya* or *Paracalamostachys*, depending on the position of attachment of the

TEXT-FIG. 10. An arborescent lycopsid cone (*Flemingites* or *Lepidostrobus*). From Thomas and Spicer (1987).

sporangia (Text-fig. 12). A cone known as *Calamocarpon*, described from petrified specimens, has only one large megaspore per sporangium but is difficult to recognize in compressions.

*Sphenophylls*
These sphenopsids were related to the calamite horsetails, but were much smaller plants, probably forming a low-lying, scrambling type of vegetation on parts of the river levees (Text-fig. 13). Their creeping stems were normally less than 10 mm wide and bore whorls of cuneate or deeply digitate leaves called *Sphenophyllum*. The shape of the leaves tended to vary quite markedly in each species, being deeply digitate on the wider stems and more entire-margined on the thinner stems. The reproductive organs, known as *Bowmanites*, were borne terminally on the stems, and were complex cones with deeply incised sporophylls and clusters of sporangia

TEXT-FIG. 11. Reconstruction of the giant horsetail *Calamites*, about 10 m tall. After Hirmer (1927).

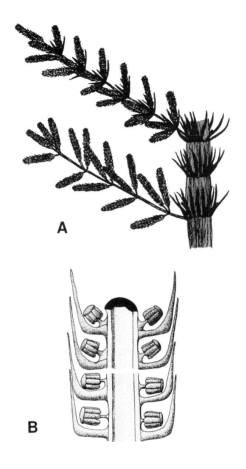

TEXT-FIG. 12. A, reconstruction of *Calamites* shoot bearing cones. B, diagrammatic longitudinal sections of *Palaeostachya* (upper) and *Calamostachys* (lower).

*Ferns*

Coal Measures ferns varied in growth-form from small, scrambling-plants to tree-ferns, but remains of the latter are generally commoner. The tree-ferns were similar in form to living species of the Cyathaceae, but belong to the order Marattiales. The reconstruction shown in Text-figure 14 is based on evidence from coal-balls, and thus represents a tree that was growing in the floodbasin swamp-forests. However, there is no reason to think that it differed markedly from the trees growing on the levees and

TEXT-FIG. 13. Partial reconstruction of *Sphenophyllum emarginatum*. After Batenburg (1977).

TEXT-FIG. 14. Reconstruction of the marattialean tree-fern *Psaronius*, about 8 m tall. After Morgan (1959).

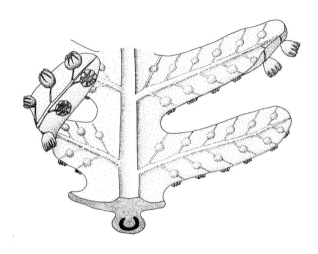

TEXT-FIG. 15. Fertile pinna segment of the marattialean fern *Scolecopteris*. From Thomas and Spicer (1987), after Millay.

represented in the adpression record. (Compare, for instance, earlier reconstructions by Corsin *in* Dalinval 1960, which were based mainly on adpression evidence.) Tree-fern trunks show few distinguishing characters, except near the crown where there are leaf-scars, the shape and distribution of which can be used to separate form-genera (*Caulopteris*, *Megaphyton*, *Artisophyton*). The fronds mostly belong to *Pecopteris*, *Cyathocarpus*, *Polymorphopteris* and *Lobatopteris*, which are distinguished on a combination of sterile and fertile characters. Most can be recognized by having more or less unlobed pinnules. The sporangia, which when found petrified are called *Scolecopteris*, *Acitheca* or *Asterotheca*, are usually in small clusters (known as sori) attached to the underside of some of the pinnules (Text-fig. 15).

The smaller, herbaceous species usually have more incised pinnules. They have a variety of fertile structures, which are used to separate the form-genera, such as *Renaultia* and *Zeilleria*. Their affinities are still a matter of some conjecture, but probably belong to various extinct families, such as the Botryopteridaceae, Zygopteridaceae, Crossothecaceae, Urnatopteridaceae (Brousmiche 1983).

TEXT-FIG. 16. Reconstruction of the trigonocarpalean gymnosperm *Medullosa noei*, about 10 m tall. After Stewart and Delevoryas (1956).

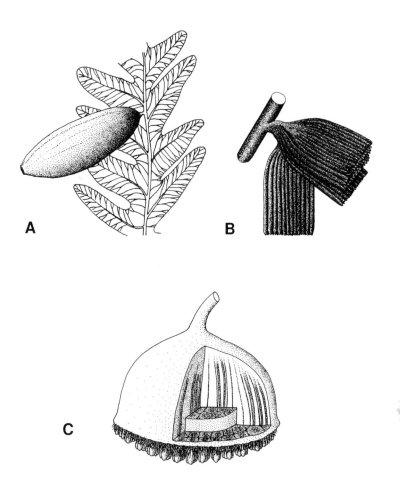

TEXT-FIG. 17. Reconstructions of trigonocarpalean fertile structures. A, the seed *Trigonocarpus*. B, the pollen-organ *Whittleseya* with a stack of flattened synangia. C, the pollen-organ *Potoniea*. B and C after Millay and Taylor (1979).

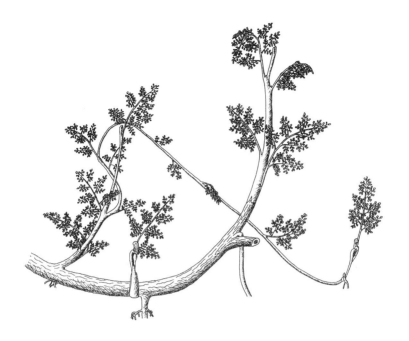

TEXT-FIG. 18. The shrubby, scrambling gymnosperm *Callistophyton* with sphenopteroid pinnules. From Thomas and Spicer (1987), after Rothwell.

*Pteridosperms*
These plants, which varied in form from small trees (Text-fig. 16) to scrambling creepers (Text-fig. 18), lived mainly on the raised levees of the rivers. Unless there is foliage attached, or some evidence of cell structure preserved, there is little to distinguish the trunk and branches of these plants. Fragments of their leaves, which superficially resemble portions of fern-fronds, are abundant fossils in the Coal Measures. There are many form-genera now used to identify them, distinguished on various features of pinnule-shape and rachis branching patterns, and including *Neuropteris*, *Alethopteris*, *Paripteris*, *Linopteris*, *Alethopteris*, *Lonchopteris*, *Eusphenopteris*, *Palmatopteris*, *Mariopteris*, *Karinopteris*, *Lyginopteris* and *Dicksonites*. Female reproductive organs were seeds attached to the fronds, and belong to form-genera such as *Trigonocarpus* (Text-fig. 17A), *Lagenospermum* and *Holcospermum*. These seeds were quite large in some species, sometimes over 100 mm long. The male organs are clusters of pollen-bearing structures, including *Whittleseya* (Text-fig. 17B) and

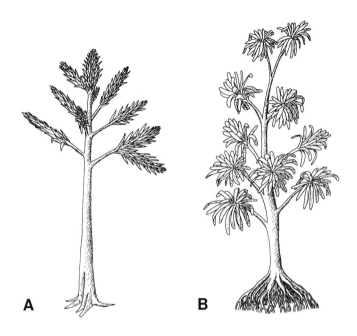

TEXT-FIG. 19. Reconstructions of cordaite trees. From Thomas and Spicer (1987). A, after D. H. Scott. B, after Cridland.

*Potoniea* (Text-fig. 17C).

The classification of this extinct group is difficult and it has even been suggested (Crane 1985) that it is not a natural group at all. Traditionally, most Coal Measures pteridosperms have been assigned to three orders: the Trigonocarpales, Lagenostomales and Callistophytales. The reconstructions given here are of one of the larger forms, belonging to the Trigonocarpales (Text-fig. 16) and a scrambling form of the Callistophytales (Text-fig. 18). Further information on these plants can be found in general reviews by Rothwell (1981), Stidd (1981) and Taylor and Millay (1981).

*Cordaites*

These were probably mainly trees up to 30 m high (Text-fig. 19), although a much smaller form has also been recently discovered (Text-fig. 20). It seems that they occupied a wide range of habitats, including the floodbasins, the levees, sea coasts and the so-called 'uplands' (i.e. extra-basinal

TEXT-FIG. 20. Reconstruction of scrambling cordaite tree. From Thomas and Spicer (1987), after Rothwell and Warner.

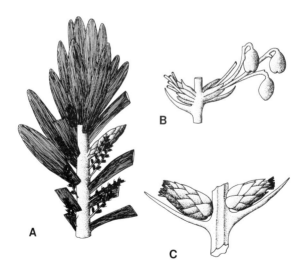

TEXT-FIG. 21. *Cordaitanthus*. A, cordaite branch with lateral fertile shoots. B, detail of fertile shoot with stalked ovules. C, detail of fertile shoot with pollen sacs. From Thomas and Spicer (1987), after Grand'Eury, Taylor and Millay, and Delevoryas.

habitats). They were primitive relatives of the conifers. The arborescent species had a straight trunk and a crown with long, strap-like leaves (*Cordaites*). Both male and female reproductive structures were cones, which when found whole are known as *Cordaitanthus* (Text-fig. 21). More usually, however, isolated seeds (*Cordaicarpus* or *Samaropsis*) are found, which are flattened and have more or less distinct lateral wings. A useful review of the group is provided by Rothwell (1988) and Trivett and Rothwell (1991).

# TAPHONOMY

Most plant fossils are only fragments (e.g. branches, leaves, reproductive organs) of the plants that once bore them. These fragments may then have been further broken up, abraded, decayed or partially eaten before becoming buried in sediment. The collector will therefore normally find isolated organs which may have been preserved some distance from where they originally grew, rather than whole plants in or near their original habitat. The process of transferring the plant fragment from the biosphere to the lithosphere is known as taphonomy, and can be divided into three essentially successive phases. For general reviews of this topic, see Spicer (1989), Bateman (1991); also various papers by Briggs, Scott and Spicer *in* Briggs and Crowther (1990).

*Producing the plant fragment*
The process of detaching the fragment from the parent plant is sometimes referred to as necrology. The detachment may either have been by abscission, as part of the normal life processes of the plant, or by a traumatic event caused by external influence.

The commonest example of natural abscission in the Coal Measures vegetation was probably the detachment of propagules—spores in the case of pteridophytic plants, seeds in gymnosperms. Normally, of course, such abscissed propagules would have germinated and thus not be preserved. However, they may well not have been 100 per cent viable, especially after landing in the sort of aqueous environment that most fossil-bearing shales represent.

Foliage abscission, which is so familiar to those of us living in temperate climates, seems to have been less common in the Late Carboniferous tropical forests. Some trigonocarpalean fronds, such as *Paripteris* Gothan, seem to have had pinnules which were readily detached, but whether this reflects abscission or simply a weak point of attachment is not clear. Certainly, no unequivocal evidence of an abscission layer has been documented in Coal Measures foliage.

It is likely that most vegetative plant macrofossils found in the Coal Measures represent fragments detached from the plant by traumatic events, as most of the compound leaves occur as irregular fragments, rather than as whole leaves or consistently as particular parts of leaves (e.g. ultimate pinnae or pinnules—the above-mentioned *Paripteris* is a possible exception). Also, when preservation is good enough, the fragments normally have a fresh appearance, as if they have been 'plucked' from the plant, rather than having withered and died. There is little evidence that animals were involved in this process. A more likely agent would have been high winds, possibly associated with the storms which generated the floods linked with the next step in the taphonomic process.

Another process of producing plant fragments in a forest situation is wildfire (Cope and Chaloner 1985). Although such fires would have destroyed much of the vegetation, the high temperatures can also result in parts of the plant being charcoalified (Scott 1989). The resulting fusain ('mineral charcoal') tends to be brittle and fragments very easily. When examined under a scanning electron microscope, however, the fragments can show exquisite preservation of cell structure (Scott and Collinson 1978; Scott and Chaloner 1983; Scott 1989).

*Transport and burial*

On being detached from the plant, the potential fossil fragment may occasionally have fallen in sediment close to the place of growth. Although rare, such fossils (autochthonous or hypoautochthonous *sensu* Bateman 1991) can be important as they tend to be relatively undamaged and can provide vital information on features such as the architecture of the larger leaves (e.g. Zodrow and Cleal 1988). Normally, however, the fragment entered the next phase of the taphonomic process, transportation and burial, or biostratinomy.

In order to determine the biostratinomic processes operating in the Late Carboniferous, attempts have been made to examine Recent analogues of such tropical forests, the most notable being by Scheihing and Pfefferkorn (1984) in the Orinoco Delta of Venezuela, and by Gastaldo *et al.* (1987) in the Mobile Delta of Alabama. Both studies showed that the plants of the levees, although forming only a small part of the overall forest vegetation, dominate the remains found preserved in the clastic sediments. They differed, however, in the exact position and mechanism of their burial. In the Mobile Delta, Gastaldo *et al.* found that the best preserved material occurred in crevasse splays, and that the fragments had been transported a considerable distance by the river waters. In the Orinoco, on the other hand, Scheihing and Pfefferkorn found plant fragments were preserved best in the distributary channels and channel-margin sediments, relatively near to where the plants grew; the interdistributary bay sediments only contained the remains of the immediately surrounding vegetation, which, in the case of the Late Carboniferous tropical forests, would presumably have been mainly calamite horsetails.

One explanation for this difference may be that, in the Mobile Delta, crevasse-splays are a significant sedimentary process in the interdistributary bays, whereas overbank flooding is the usual response to high-water events in the Orinoco. Fielding (1984*a*) has argued that most of the Coal Measures floodbasin sediments were generated by crevasse-splays rather than overbank-flooding, which would make the Alabama Mobile

Delta the best Recent analogue. If so, it would seem likely that most of the Coal Measures plant fossils preserved in shales and mudstones were transported by river water for some considerable distance from their place of growth, and were dumped into the interdistributary bays through crevasses in the levee-banks. However, much more work needs to be done on this problem; it is unlikely that this was the only biostratinomic process in operation at the time.

*Post-burial alteration*
When the plant fragment finally found its way into the sediment, the last taphonomic phase commences—diagenesis. This alters the shape, structure and composition of the fragment, changing it into the fossil that we eventually find in the rock. Since the diagenetic processes can be quite varied, they can result in a number of different preservational states.

The classification of preservational states has been the subject of controversy, and different schemes have been proposed. The most recent is by Bateman (1991), who defines different classes of preservational state on two major factors: (1) the degree of compression that has occurred; and (2) the composition of the remaining organic components. The essence of the classification is summarized in Table 1 and Text-figure 22.

Anatomically-preserved petrifactions are well known from the Coal Measures, such as the permineralizations found in coal-balls (Phillips 1980, Scott and Rex 1985). The fusain fragments produced by pre-burial forest wildfire can also yield fine anatomical detail (see above). However,

TABLE 1. Classification of main preservational types of plant fossils, after Bateman (1991).

|  | Organic component remaining | | |
| --- | --- | --- | --- |
|  | Carbon plus volatiles (Volatilized) | Carbon only (Devolatized) | None (Decarbonized) |
| Petrifaction (<80% compression) | Permineralization | Fusain | Petrifaction (auct.) |
|  |  |  | Mould/cast/ authigenic replica |
| Adpression (>80% compression) | Compression | | Impression |

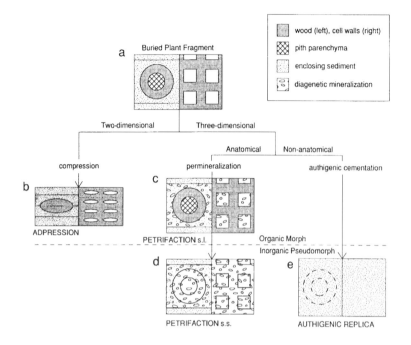

TEXT-FIG. 22. Summary of the modes of preservation of plant fossils. From Bateman (1991, fig. 2.2).

they have to be studied in the laboratory using quite different methods, and will not be further dealt with here. The fossils covered by the key are either adpressions or decarbonized petrifactions.

*Adpressions.* After the plant fragment was buried in anaerobic, waterlogged sediment, decay usually ceased. Further accumulation of overlying sediment then caused vertical compaction both of sediment and organic tissues. The plant fragments thereby lost some or all of their volatile organic components, and become converted into coal. This coalification destroyed most of the cellular structure, and distorts the shape of the plant fragment, mostly in a vertical direction (Rex and Chaloner 1983; Thomas 1987*a*). The resulting thin, black or dark brown layer on the rock is known as a phytoleim. If the phytoleim was lost, either by oxidation of the sediment after compaction, or by removal on splitting the rock, then an impression (or decarbonized adpression) of the plant fragment will be embedded in the rock surface.

Splitting a rock along the plane of an adpression fossil results in two fossils, a *part* and a *counterpart*. If the phytoleim is still preserved, the part and counterpart reflect the same surface of the plant, but in specimens where all organic remains have been lost, the impression pairs represent different surfaces and may thus show different features.

Such adpressions can be studied by observation either by light microscopy or scanning electron microscopy. If 28 per cent or more of the volatile components is still present, the phytoleim can also be treated with oxidative acid to remove the coaly substance, liberating any epidermal cuticles or spore coats still preserved for microscopic study (Cleal and Zodrow 1989; Cleal and Shute 1991, 1992). Decarbonized adpressions will show the form of the plant fragments and, in some cases, will yield fine details of the plant surface (Chaloner and Collinson 1975*a*).

*Authigenic replicas.* If mineralization occurs in the sediment before significant compression has occurred, then any plant fragments that it contains will be able partly to avoid the distorting effects of subsequent compression. In ironstone nodules, this mineralization does not permeate the fabric of the cell tissue (i.e. permineralize it), but merely forms a replica of its outer, three-dimensional morphology. The plant tissue will subsequently be changed to carbon. Sometimes, this is lost, resulting in a hollow mould. Alternatively, subsequent mineralization fills the voids. The classic examples of such plant fossils in the Upper Carboniferous are found at Mazon Creek (Darrah 1969), but there are also British examples such as those found (but alas no longer) at Coseley near Dudley (Arber 1916). Such fossils can preserve the morphology of fine structures such as fern sporangia in remarkable detail. They are studied mainly in the same way as adpressions.

*Casts and moulds.* Structurally robust plant material, such as woody stems, may not be severely distorted by sediment compaction. In such cases, when the sediment becomes lithified, it forms a three-dimensional mould of the plant fragment. Normally, the plant tissue would rot away and so the mould is preserved as a hollow cavity in the rock. If the mould becomes secondarily infilled by sediment, the result is known as a cast.

Casts can also form when part of the plant was hollow in life, or if part of it consisted of very delicate tissue that degraded immediately after death, producing a cavity which could be filled with sediment. The best known examples of this are the stems of cordaite trees (*Artisia*) and the giant horsetail *Calamites*, both of which had a central core of easily degradable pith, surrounded by wood; pith casts of *Calamites* are in fact more commonly found than casts or adpressions of the whole stem. Such casts often occur just as sediment fills. Sometimes, however, they also have an outer layer of carbon, representing the remains of the more robust tissue that became compressed between the sediment cast and the rock matrix. In such cases, the one fossil may preserve internal features of the plant (e.g. the internal surface of the woody cylinder of a calamite stem) as well as the outer surface details (Thomas 1987a; Leary and Thomas 1989).

Casts and moulds rarely show fine details of the plant, and in some cases it is difficult to determine even their general affinities. The most useful examples are the *in situ* stumps of trees preserved as moulds or casts, since they provide evidence as to the spatial distribution of the trees within the forest and, in some cases, the relative distributions of different plant groups (e.g. Gastaldo 1987).

*Taphonomic filtration*
The processes discussed above clearly bias the sample of the Late Carboniferous tropical forests preserved in the fossil record; this biasing may be referred to as taphonomic filtration. The following factors control the filtration, allowing a particular plant structure to become preserved.

1. How many of the particular organ were produced by each plant;
2. Whether or not the organ was naturally shed;
3. Rate of loss of the organ;
4. The size of the plant relative to the adjacent vegetation;
5. Durability of the organ;
6. Degree of animal predation and fungal attack;
7. Specific gravity of the organ;
8. Rate at which the organ became waterlogged;
9. Where the plant grew.

Factors 1 to 4 will influence the number of specimens which are available to become fossils, and factors 5 to 9 will influence the probability that they will find their way into suitable sediment for fossilization.

To summarize, the plant fossils represented in this key are dominated by the remains of the larger plants (mainly trees and shrubs) that grew on or near the levees in the Late Carboniferous tropical forests. Remains of herbaceous plants are found, but in much smaller number. It may be that such plants were just less abundant, but it could also be a consequence of such plants producing fewer leaves, propagules, etc. Also, the effect of baffling by nearby trees may have partially protected them from wind-damage, and thus restricted the number of fragments from them that entered the taphonomic chain. Among the plants that were initially sampled, taphonomic filtration would tend further to bias the fossil composition in favour of the more robust leaves, seeds and stems. Delicate fructifications, such as gymnospermous microsporangiate organs, tend to be very rare, even though they must have been abundantly produced by the plant.

# COLLECTION AND CONSERVATION OF SPECIMENS

Plant macrofossils occur throughout the British Coal Measures, even in the marine bands. However, the most diverse and well-preserved assemblages are from the mudstones and siltstones of the floodbasins. Most specimens found here are transported (and therefore mainly fragmentary) pieces of leaves and reproductive organs. The large, divided leaves of the ferns and pteridosperms could not travel far before breaking up, and so more complete specimens occur in strata associated with the levee banks (e.g. fluvial slump deposits, crevasse channel fills). In contrast, the more robust stems could withstand the higher energy conditions of the distributary channels and thus may be found in fluvial sandstones. Stems could also float further than leaf fragments and so may occur in marine shales.

The fossil-bearing shales are relatively soft, and are best dealt with using a 0·5 kg geological hammer and small cold-chisels. Each specimen collected from these shales should be individually wrapped in absorbent paper (e.g. old newspaper) to protect the surface of the fossil, which may show important fine detail. If possible, both the part and counterpart of the fossil should be collected. Each specimen should be carefully labelled in the field with locality, horizon and any other relevant details.

Collecting specimens from the harder sandstones requires heavier equipment, such as a 1 kg geological hammer, larger cold chisels and a bolster chisel. They may not be so well preserved as those from the mudstones and shales, but can often be significantly larger, showing the overall form of the plant better (cf. Scott 1978, pl. 27 fig. 1; Zodrow and Cleal 1988, pl. 2). Such sandstones tend to be poorly laminated, and so the fossils often break up when being extracted. Usually, the only solution is to break the fossil into as few pieces as possible, and then to glue them back together at home.

On returning home, the fossils should be allowed to dry slowly (never apply heat); many roof shales will crumble if dried too quickly. The surfaces of specimens should never be coated with varnish, since this can damage important surface detail. Broken specimens may be glued back together with a strong adhesive, but it should be kept off the fossil surface. Specimens should then be stored, preferably one to a box or cardboard tray, and labelled. As much information as possible should be included on the label; the absolute minimum should be details of the exact locality and horizon (if known) from where it originated.

# FOSSIL LOCALITIES

Coal Measures plant fossils can be collected from many different types of locality in Britain.

*Coastal exposures.* These are potentially the best natural exposures for collecting plant fossils. Where there is extensive foreshore outcrop, large slabs of fossiliferous rock should be readily accessible. It should also be easy to determine details of the stratigraphy and sedimentology of the fossil horizon. Weathering by sea water and the atmosphere can be a problem, and the collector will often have to hammer well into the rock to obtain good specimens. Collecting from cliff faces should normally be avoided, and any exposure near the cliff should only be approached when wearing a 'hard hat'. Attention should also be paid to the times of high tides.

*Inland natural exposures.* These are mainly in stream and river sections, and should be easily accessible, providing prior permission is obtained from the landowner. Most such exposures are, however, quite small and only yield small specimens. The most productive outcrops are in stream beds, but the specimens there have to be taken from below water. Weathering may be a problem at these sites, and well-preserved material may be difficult to find. Natural scarps bearing plant fossils are rare in Britain. Where they do occur, the same care should be taken as when visiting coastal cliffs.

*Working quarries, claypits and opencast coal mines.* These are some of the best localities for Coal Measures plant fossils. Exposure is usually good and the working rock faces will be continually uncovering fresh material. Large slabs with well-preserved fossils can often be found, and details of the stratigraphy and sedimentology observed with comparative ease. There is always danger from plant machinery and falling rocks, and 'hard hats' must be worn. Written permission should be obtained from the developer in advance of a visit.

*Disused quarries, claypits and opencast coal mines.* Abandoned workings are becoming scarce in Britain, because of their value as waste disposal sites. They do not have plant machinery, but there may be danger from rock falls from the weathering face. This weathering also means that the fossils are often not well preserved. A 'hard hat' should always be worn.

*Underground coal mines.* Collecting from underground workings is rarely possible nowadays because of the mechanized working of the face; few colliery managers will allow inexperienced people below ground, especially if they expect to be hammering in the workings! Organized group visits can sometimes be arranged, but this rarely provides satisfactory opportunities to collect fossils. In older, less mechanized mines it is sometimes possible to collect from near the face, while miners are digging the coal. From personal experience, however, the authors can vouch that this can be dangerous, and is generally not recommended.

*Colliery spoil tips.* Modern mechanized collieries usually only produce very fragmented spoil containing only small plant fossils. Tips of older mines, however, often yield much larger plant fossils. The material is not *in situ*, and so the stratigraphy and sedimentology cannot be observed directly; but it may be possible to find out which seams were worked by the mine, providing a rough guide to their stratigraphical origin. Tips are relatively safe to collect from, although care should be taken on the uneven surface. A 'hard hat' should be worn, particularly on the steeper tips. If the colliery is still working, permission must be obtained from the colliery manager.

There are far too many Coal Measures plant localities in Britain to give a detailed list. Furthermore, many of the best are temporary exposures with a short life expectancy. Some of the classic localities are mentioned in the biostratigraphy section towards the end of this volume, while others are detailed in published excursion guides (Neves and Downie 1967, Bassett and Bassett 1971, Savage 1977, Bassett 1982, Austin *et al.* 1985) and the guide to 'British Fossiliferous Localities' published by the Palaeontographical Society. Detailed accounts of the geology of the British coalfields, including the localities yielding plant fossils, can be found in the relevant sheet memoirs of the British Geological Survey. More general accounts can be found in the *British Regional Geology* series also published by the BGS, and in the Geological Society volumes by Craig (1991) and Duff and Smith (1992).

# INTRODUCTION TO THE KEYS

*Nomenclature and classification*
As most plant fossils are only isolated fragments of plants, palaeobotanists have developed a system of taxonomic classification that is different from that used for living plants. Although it uses a similar style of Linnean binomial nomenclature, it differs in that a particular taxon usually refers only to a particular organ or organ aggregate; only in the case of the fossils of some small, herbaceous plants (e.g. the fern *Nemejcopteris* Barthel, 1968) has it proved practical to use a taxon for an entire plant. For instance, when a specimen is identified as *Flemingites*, it is implicit that it is a lycopsid cone. There is no such thing as a *Flemingites* leaf or stem, even though the plant that originally bore the cone obviously had leaves and stems (in this case they would be identified as *Cyperites* and *Lepidodendron*, respectively). In order to distinguish such names from whole-plant species and genera, they are usually referred to as form-species and form-genera (Cleal 1986*a*; Thomas *in* Briggs and Crowther 1990). At first glance, this system may seem to produce unnecessarily cumbersome and highly artificial lists of names; a list of a dozen taxa could in fact be generated by only three or four original plants. However, it provides a means of communication which not only gives a fossil a name, but also provides information about the kind of organ that it is based on. The problems of communication would be greater, not less, if form-genera could include different organs through synonymy.

Occasionally, organic attachment can be found between organs that previously received separate names, proving that they came from the same parent plant. Consistent association can also indicate original botanical unity, but such evidence is merely circumstantial and not proof. Even after demonstrating the organic attachment between two or more organs, the form-taxa are not abandoned by consigning them to a synonymy (Chaloner 1986). Divergent or convergent evolution may have resulted in two or more different plants having, for example, similar shoots but different reproductive organs. Also, differing rates of evolution within plants could alter one of their organs beyond recognition, so that a form-genus of shoots might be found to overlap stratigraphically with the separate distributions of more than one form-genus of its reproductive organs.

*Division of Keys*
Since each of the major organs of the plants are given their own names, it has proved more practical to divide the identification key into sections

based on morphology rather than on biological affinity. The division adopted is essentially that used by Chaloner and Collinson (1975*b*).

Group A: Stems or stem-like fossils with longitudinal striations or ribs, but without regularly distributed leaf scars. The group includes pith casts and roots of calamite horsetails; and leaves, stems and roots of cordaites.

Group B: Stems with leaf scars distributed regularly in vertical rows or spirally around the stem. They mostly belong to lycopsids, but also includes some ferns and pteridosperms.

Group C: Leaves borne in whorls around narrow stems, belonging to calamites and sphenophylls.

Group D: Branching stems covered with small, tapered leaves, belonging to lycopsids and conifers.

Group E: Large fern and pteridosperm fronds or parts of such fronds (pinnae and pinnules).

Group F: Cigar-shaped 'cones' or strobili borne singly or in a series along a stem. They have overlapping scales arranged around an axis, and may belong either to lycopsids, horsetails or cordaites.

Group G: Detached lycopsid cone scales or leaves. Triangular or more or less parallel-sided and tapering at the apex, with a single central vein. A swelling (sporangium) may be present at the base.

Group H: Seeds and bell-shaped clusters of pollen sacs. Most belong to pteridosperms, but the group also includes isolated cordaite seeds and some clusters of fern sporangia.

# KEY TO GROUP A: STEMS OR STEM-LIKE FOSSILS WITH LONGITUDINAL STRIATIONS OR RIBS

This is a mixed group of stems and leaf remains. The majority are stems of large sphenopsids ('calamites'), preserved either as adpressions of the outer surfaces, or of the inner surface surrounding the central, hollow pith cavity. Some may be casts of the pith cavity. Large *Calamites* stems are very rarely found with leaves still attached, and the terminal leafy shoots will key out in Group C. All calamite stems and pith casts show longitudinal striations, or deep grooves, broken horizontally at intervals by the nodes marking the regularly spaced zones that bore the leaves and branches. Circular or oval or oval-elongate scars, called infranodal canals, may be seen at the upper ends of pith cast ribs marking the former positions of vascular rays. The portions between the nodes are called internodes. *Calamites* species are distinguished on a combination of nodal and internodal characters (e.g. Crookall 1969). Fragments, therefore, may be impossible to identify.

The strap-shaped leaves, called *Cordaites*, are from the gymnosperm tree generally referred to informally by the same name. They are distinguished on outline and on the veining pattern (e.g. Crookall 1970). Unfortunately, they are usually found in a fragmentary condition and only very rarely attached to shoots. The pith casts of *Cordaites* stems are called *Artisia*.

1. Longitudinal striations or ribs divided into segments by transverse lines marking the positions of nodes: stems..........................................2
Longitudinal striations or ribs not divided into segments by transverse lines: leaves..............................................................11
2. Transverse lines usually distant. Longitudinal striations always continuous throughout each segment...................................................3
Transverse lines close. Longitudinal striations not always continuous throughout each segment ............................................................*Artisia*
(Text-fig. 40B; pith casts of cordaite stems which cannot be classified into meaningful species)

---

EXPLANATION OF PLATE 1

Fig. 1. *Cordaites principalis*. BMNH V.29275; Coal Measures (Westphalian); Sunderland, × 0·7.
Fig. 2. *Calamites suckowii*. NMW 92.20G2; Six Quarters Coal (upper Langsettian); Wythemoor House Opencast site, Cumbria, × 1.

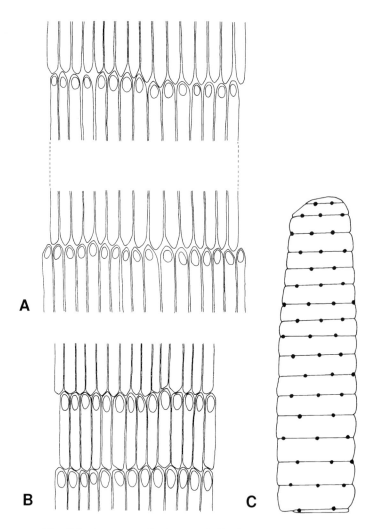

TEXT-FIG. 23. *Calamites* stems. A, *C. cistii*, indicating variable nodel length, × 2. B. *C. suckowii*, × 2. C. *C. multiramis*, showing relative positions of nodal branch scars, × 0·25. After Crookall (1969).

---

TEXT-FIG. 24. *Calamites carinatus*. A, stem showing three nodal branch scars, × 1. B, schematic representation of relative positions of opposite pairs of nodal leaf scars, × 0·25. After Crookall (1969).

## Key to Group A

51

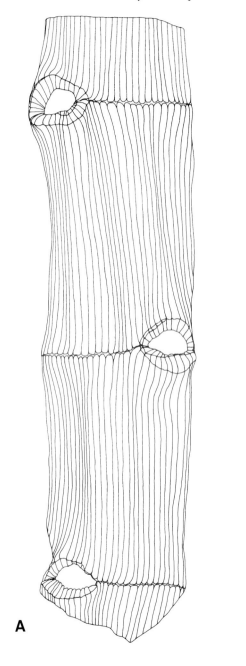

A

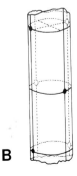

B

3. Axes less than 10 mm broad. Leaves attached to the nodes seen at the sides of the stem ............................................................................ See Group C
   Axes more than 10 mm broad. Leaves only very rarely attached to the nodes ..................................................................................................... 4
4. No large round/oval branch scars visible at the nodes or only very rarely present ........................................................................................ 5
   Large round/oval nodal branch scars regularly present .................... 6
5. Internodes longer than broad, tubercles oval-elongate never as wide as the ribs ................................................................... *Calamites cistii*
   (Text-fig. 23A)
   Internodes broader than long, tubercles as wide (or almost) as the ribs ...................................................................... *Calamites suckowii*
   (Pl. 1, fig. 2; Text-fig. 23B)
6. Branch scars present on every node, or separated by an occasional branchless node ..................................................................................... 7
   Branch scars not on every node, separated by more than one branchless node ..................................................................................... 9
7. Branch scars in opposite alternating pairs, ribs convergent towards branch scars ........................................................ *Calamites carinatus*
   (Text-fig. 24)
   Branch scars more than two per node, ribs not converging towards branch scars ......................................................................................... 8
8. Branch scars regularly alternating. Internodes variable in length, short internodes may be single or in groups that are generally separated by one long internode ............................................ *Calamites multiramis*
   (Pl. 2, fig. 1; Text-fig. 23C)
   Branch scars not regularly alternating. Internodes all approximately the same length, every seventh or eighth node without a branch scar ..
   ........................................................................... *Calamites brongniartii*
   (Text-fig. 25D)
9. Internode lengths uniform between branch-scar nodes, branch-scars circular/oblong to rectangular, close together and forming a complete ring ..................................................................... *Calamites goeppertii*
   (Pl. 2, fig. 2; Text-fig. 25A)
   Internode lengths not uniform between branch-scar nodes (the shortest internode above and the longest below), branch scars circular, distant and not forming a complete ring ....................................................... 10

---

EXPLANATION OF PLATE 2

Fig. 1. *Calamites multiramis*. NMW 90.9G1; Farrington Formation (upper Westphalian D); Kilmersdon Colliery Tip, Radstock, Avon, × 1.

Fig. 2. *Calamites goeppertii*. NMW 90.9G2; Farrington Formation (upper Westphalian D); Kilmersdon Colliery Tip, Radstock, Avon, × 1.

## Key to Group A

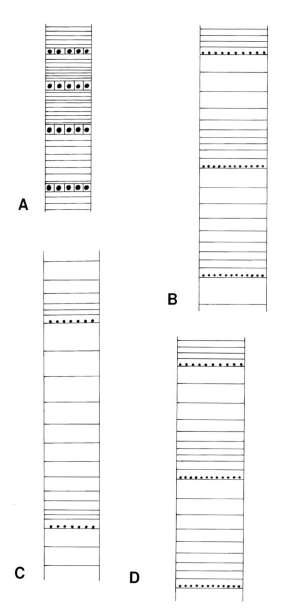

TEXT-FIG. 25. Schematic drawings showing relative positions of nodal leaf scars in *Calamites*. A, *C. goeppertii*. B, *C. schuetzeiformis*. C, *C. undulatus*. D, *C. brongniartii*. A, B, D after Crookall (1969), C after Boureau (1964). All × 0·25.

## Key to Group A

10. Internode lengths often irregular, vertical ribs distinct ..........................
................................................................*Calamites schuetzeiformis*
   (Text-fig. 25B)
   Internode length regular, vertical ribs only slightly raised and often indistinct ............................................................*Calamites undulatus*
   (Text-fig. 25C)
11. Leaves narrower than 4 mm, with a very few longitudinal ridges and furrows ....................................................................................................12
   Leaves broader than 4 mm, with many longitudinal lines ...................13
12. Leaf margins entire ..............................................*Cyperites bicarinatus*
   (Pl. 5, fig. 3; Text-fig. 26B)
   Leaf margins ciliated ...................................................*Cyperites ciliatus*
   (Text-fig. 26A)
13. All lines (veins) of equal visibility and size .........................................14
   Thick lines (veins) separated by thinner, less distinct lines (veins)...15
14. Leaves 4–8 mm broad, apex obtuse .................*Cordaites microstachys*
   (Text-fig. 27C)
   Leaves 5–100 mm broad, apex acute or bluntly pointed .......................
   ......................................................................*Cordaites palmaeformis*
   (Text-fig. 27C)
15. Thicker veins alternating with thinner veins .......................................16
   Thicker veins separated by two or more veins ...................................17
16. Leaves less than 10 mm broad ........................*Cordaites microstachys*
   (Text-fig. 27D)

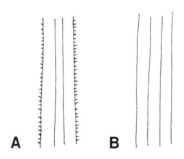

TEXT-FIG. 26. Portions of *Cyperites* leaves (leaf topographies may vary considerably). A, *C. ciliatus*. B, *C. bicarinatus*. A after Crookall (1964), B after Rex (1983). Both × 2.

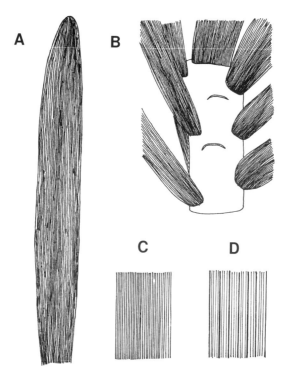

TEXT-FIG. 27. *Cordaites* leaves. A, single leaf of *C. principalis*, × 0·2. B, portion of leafy shoot of *C. angulosostriatus*, × 0·2. C, examples of venation pattern with all veins of equal visibility and size, × 10. D, examples of venation pattern with thick veins separated by thinner and less distinct veins, × 10. B after Grand'Eury (1877).

    Leaves 30–120 mm broad ................................*Cordaites borassifolius*
                                                            (Text-fig. 27D)
17. Leaves 30–60 mm broad, apex rounded, 1–6 (usually 2–3) thinnner veins between thicker veins ................................*Cordaites principalis*
                                                     (Pl. 1, fig. 1; Text-fig. 27A,D)
    Leaves 40–120 mm broad, apex bluntly pointed, 2–5 thinner veins between the thicker veins ...........................*Cordaites angulosostriatus*
                                                          (Text-fig. 27B,D)

# KEY TO GROUP B:
# AXES WITH SCARS DISTRIBUTED REGULARLY IN VERTICAL ROWS OR SPIRALS

These are stems or stem-like rooting organs bearing regularly spaced scars that mark the dehiscence points of whole leaves, leafy parts or roots. Most are stems of large arborescent lycopsids that grew in an unusual manner. The main axis, the trunk, soon developed a massive apex which, after growing to a height of anything up to 40 m, subsequently dichotomized to give smaller and smaller apices. As the stems aged they shed the distal parts of their leaves, the laminae, often leaving the swollen basal part attached to the stem. These swollen bases, which were most probably photosynthetic, are either distinct and in recognizable spirals (*Lepidodendron*, *Lepidophloios*, *Sublepidophloios* and some species of *Sigillaria* and *Bothrodendron*) or fused into vertical ribs (most species of *Sigillaria*). Those stems with leaves still attached are usually the more terminal, narrower twigs. They are dealt with in Group D.

The distinct swollen bases are called leaf cushions (Text-fig. 9). The larger stems have larger leaf cushions which results in a considerable amount of size variation within a species. The diameter of the stem, and the size of the leaves or of the leaf scars therefore reflect the position on the parent plant rather than age. A limited amount of secondary thickening within the stems can also cause the cushions to separate in both *Lepidodendron* (Text-fig. 8) and *Sigillaria*. Other variations in growth pattern in *Sigillaria* can even result in different shaped leaf scars (Thomas 1972). Large amounts of secondary thickening in the main trunk can ultimately cause the loss of the outermost tissues (decortication) leaving the stem surface showing different features, that cannot be related to the species concerned (Thomas and Watson 1976). One of the commonest decorticated forms of *Sigillaria* is known as *Syringodendron*.

The scars representing the areas of leaf abscission (Text-fig. 9B) usually show a small mark of the leaf conducting tissue (vascular trace) flanked by two marks of aerating tissue (parichnos). Another scar above the leaf scar marks the aperture of a pit, which originally housed a membranous flap of tissue called a ligule. In some species of *Lepidodendron*, two further parichnos marks are visible on the cushion surface below the leaf scar (Text-fig. 32D).

Species distinction is usually through the recognition of morphological characters of the cushions and scars (Chaloner 1967; Crookall 1964,

1966), although epidermal characters have been shown to be valuable and somtimes even indispensible (Chaloner and Collinson 1975a; Thomas 1966, 1970a, 1977; Thomas and Masarati 1982).

1. Rhomboidal, hexagonal or pyriform leaf scars < 20 mm long, and with one or three point-like or short-linear markings............................2
Rounded scars with one central mark ................................................60
Oval to subrectangular leaf scars > 20 mm long, and with elongate, curved markings ..................................................................................63
Pairs of longitudinally elongated scars........................*Syringodendron*
(Text-fig. 28A)
2. Leaf scars situated on raised rhomboidal, diamond or hexagonal-shaped leaf cushions, or vertically aligned ribs .................................3
Leaf scars situated directly on the stem and not on raised leaf cushions or vertically aligned ribs ...................................................................56
3. Leaf scars on rhomboidal/diamond or teardrop-shaped leaf cushions, arranged spirally and not in vertical files ..........................................4
Leaf scars in vertical files, on hexagonal-shaped leaf cushions or vertically aligned ribs.......................................................................26
4. Leaf cushions slightly larger than leaf scars; sometimes indistinct. Inter-cushion bark usually showing straight or slightly flexuose striae running between diagonally adjacent cushions ....................................
.................................................................*Asolanus camptotaenia*
(Text-fig. 28B)
Leaf cushions much larger than leaf scars. Inter-cushion bark (if present) with roughly longitudinally running striae ............................5
5. Leaf scars on the upper half of vertically elongated leaf cushions. Leaf cushions raised/projecting, but never bulging and downturned ..........6
Leaf scars at the lower end of bulging and downturned leaf cushions ............................................................................................................23
6. Leaf cushions adjacent or nearly adjacent, not separated by wide bands of inter-cushion bark................................................................7
Leaf cushions separated by wide bands of inter-cushion bark...........20
7. Leaf cushions with rounded upper angles ...........................................8
Leaf cushions with acute/pointed upper angles..................................9
8. Leaf cushion surface covered with fine granulations ..........................
.............................................................................*Sigillaria macmurtriei*
(Text-fig. 37C)

---

EXPLANATION OF PLATE 3

*Lepidodendron aculeatum*. NMW 90.8G1; Farrington Formation (upper Westphalian D); Lower Writhlington Colliery Tip, Radstock, Avon, × 1.

TEXT-FIG. 28. A, *Syringodendron* sp.; NMW 92.20G4; horizon not recorded; Frickley Colliery, Yorkshire. B, *Asolanus camptotaenia*; David Davies Collection, NMW 22.114G630; Yard Coal (Langsettian); Mid-Glamorgan. Both × 1.

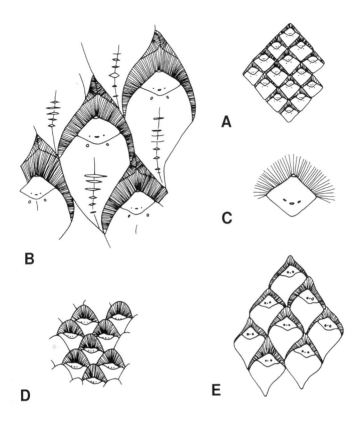

TEXT-FIG. 29. *Lepidodendron* leaf cushions, with shading lines indicating the areas covered by surface striations rather than individual striations. A, *L. dichotomum*. B, *L. mannabachense*. C, *L. mannabachense*, with ligule pit aperture adjacent to leaf scar. D, *L. peachii*. E, *L. barnsleyense*. After Thomas (1970a). All × 2.

  Leaf cushion above leaf scar 'hood-like', fine striations running from the scar to the cushion edge ............................. *Lepidodendron peachii* (Text-fig. 29D)
9. Leaf cushions with fine striations above leaf scar, running from leaf scar to edge of cushion ................................................. 10
  No fine striations above the leaf scars ................................................. 12
10. Cushions approximately as broad as long .*Lepidodendron dichotomum* (Pl. 4, fig. 2; Text-fig. 29A)
  Cushions longer than broad ................................................................. 11

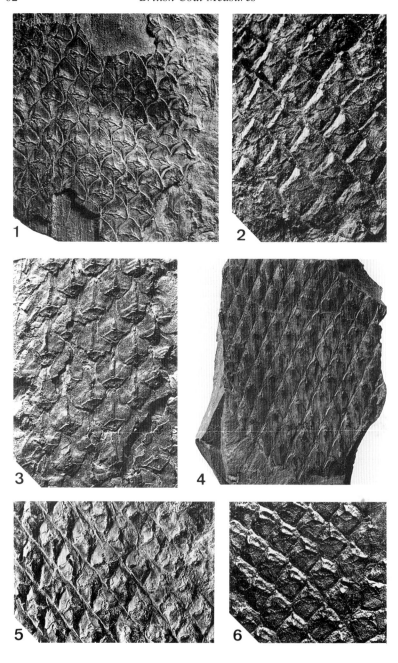

11. Upper cushion angle obtuse-rounded. External parichnos present ......
    ................................................*Lepidodendron mannabachense*
    (Pl. 4, fig. 6; Pl. 5, fig. 2; Text-fig. 29B,C)
    Upper cushion angle obtuse-pointed. External parichnos absent ...........
    ................................................*Lepidodendron barnsleyense*
    (Text-fig. 29E)
12. Leaf cushion with lines running from lateral angles of leaf scar to cushion edges ............................................................................... 13
    Leaf cushions with no lateral lines ............ *Lepidodendron subdichotum*
    (Text-fig. 30B)
13. Leaf cushions elevated above the general cushion surface on small projections (often torn off during the splitting of the rock) ................ 14
    Leaf scars level with the general cushion surface .......................... 15
14. Leaf cushions 4–5 times longer than broad. Lateral lines distinct but not raised ............................................... *Lepidodendron jaraczewskii*
    (Pl. 4, fig. 1; Text-fig. 31B)
    Leaf cushions 2–3 times as long as broad. Lateral lines distinct and raised ......................................................... *Lepidodendron feistmantelii*
    (Pl. 4, fig. 5; Text-fig. 30C,D)
15. Leaf cushions with smooth surfaces. Ligule pit apertures separated and a little distance above the upper angles of the leaf scars ................
    ....................................................................*Lepidophloios acerosus*
    (Pl. 4, fig. 3; Text-fig. 34A,B)
    Leaf cushions with transverse striations. Ligule pit aperture adjacent to the upper angle of the scar ............................................................ 16
16. Leaf cushions with very distinct striations crossing nearly the entire cushion width ...................................................................................... 17
    Leaf cushions with short striations confined to the central line of the cushions ............................................................................................. 18

---

EXPLANATION OF PLATE 4

Fig. 1. *Lepidodendron jaraczewskii*. NMW 90.20G9; Coal Measures (Westphalian); Snowdown Colliery tip, Kent, × 1.

Fig. 2. *Lepidodendron dichotomum*. Tolson Memorial Museum, Huddersfield, KLMUS:1993.299; Coal Measures (Westphalian); Huddersfield, West Yorkshire, × 4.

Fig. 3. *Lepidophloios acerosus*. BGS, Kidston Collection, 4947; Parkgate Coal (Langsettian); Hartley Bank Colliery, Horbury, West Yorkshire, × 1.

Fig. 4. *Lepidodendron lycopodioides*. NMW 75.46G601; Gwsgol Coal (Bolsovian); Gws Cwm Colliery, Burry Port, near Llanelli, Dyfed, × 1.

Fig. 5. *Lepidodendron feistmantelii*. BGS 77179. Fenton Coal (Langsettian); Wolley Colliery, near Barnsley, South Yorkshire, × 1.

Fig. 6. *Lepidodendron mannabachense*. BGS, Kidston Collection, 2469; Lower Main Coal (Duckmantian); Cramlington, Northumberland, × 2.

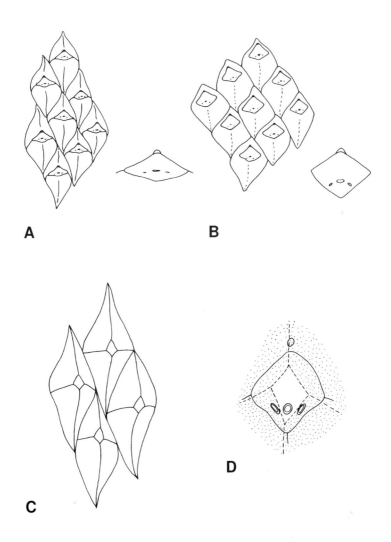

TEXT-FIG. 30. *Lepidodendron* leaf cushions. A, *L. arberi*. B, *L. subdichotum*. C, *L. feistmantelii*. D, *L. feistmantelii*, showing elevated leaf scar overlapping part of leaf cushion. After Thomas (1970a). All × 2, with enlargements of leaf scars and ligule pit apertures, × 6.

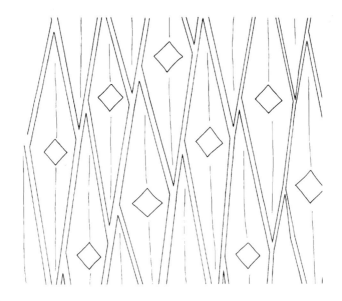

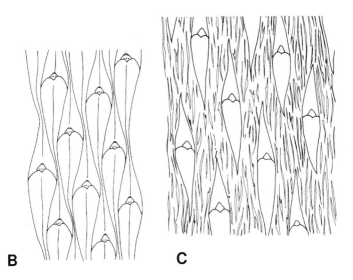

TEXT-FIG. 31. *Lepidodendron* leaf cushions. A, *L. fusiforme*, × 2. B, *L. jaraczewskii*, × 1. C, *L. rimosum*, × 1. After Crookall (1964).

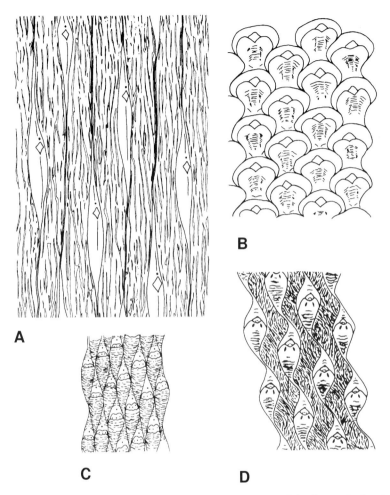

TEXT-FIG. 32. *Lepidodendron* leaf cushions. A, *L. wedekindii*, × 2. B, *L. volkmanianum*, × 1. C, *L. worthenii*, × 1. D, *L. serpentigerum*, × 1. A-C after Crookall (1964), D after Thomas (1970a).

EXPLANATION OF PLATE 5

Fig. 1. *Lepidodendron rimosum*. NMW 92.20G8; Parkgate Coal (Langsettian); Church Lane Colliery, Barnsley, South Yorkshire, × 1.

Fig. 2. *Lepidodendron mannabachense*. NMW 92.20G12; Coal Measures (Westphalian); Snowdown Colliery tip, Kent, × 1.

Fig. 3. *Cyperites bicarinatus*. NMW 90.9G5; Farrington Formation (Westphalian D); Kilmersdon Colliery tip, near Radstock, Avon, × 1.

17  Leaf cushions with no definite keel, striations present above and below the leaf scars ..................................... *Lepidodendron worthenii*
(Pl. 6, fig. 1; Text-fig. 32C)
Leaf cushions with a definite keel, striations present only below the leaf scars ................................................. *Lepidodendron volkmanianum*
(Text-fig. 32B)
18. Leaf cushions broad-rhomboidal with inflexed apical and basal angles ................................................................ *Lepidodendron aculeatum*
(Pl. 3; Text-fig. 33A,C,D)
Leaf cushions narrow-rhomboidal with straight apical and basal angles ................................................................................................ 19
19. Leaf cushions approximately twice as long as broad. Leaf scars approximately twice as broad as long ................. *Lepidodendron arberi*
(Text-fig. 30A)
Leaf cushions approximately five times longer than broad. Leaf scars approximately as long as broad ..................... *Lepidodendron fusiforme*
(Text-fig. 31A)
20. Leaf cushions with inflexed upper and lower angles ...................... 21
Leaf cushions straight, more or less symmetrical ................................
................................................................ *Lepidodendron rimosum*
(Pl. 5, fig. 1; Text-fig. 31C)
21. Leaf cushions > 15 mm long, 2–4 times longer than broad ............... 22
Leaf cushions < 15 mm long, 6–7 times longer than broad ..................
.................................................................. *Lepidodendron wedekindii*
(Text-fig. 32A)
22. Leaf cushions slightly inflexed, separated or joined by short intercushion connections .................................... *Lepidodendron aculeatum*
(Pl. 3; Text-fig. 33)
Leaf cushions distinctly inflexed, separated or joined by long intercushion connections ...................... *Lepidodendron serpentigerum*
(Text-fig. 32D)

---

EXPLANATION OF PLATE 6

Fig. 1. *Lepidodendron worthenii*. NMW 90.9G6; Farrington Formation (Westphalian D); Kilmersdon Colliery tip, near Radstock, Avon, × 1.

Fig. 2. *Lepidophloios laricinus*. BMNH V.4466B; Main Coal (Duckmantian); Newbiggin Colliery, Northumberland, × 2.

Fig. 3. *Lepidophloios macrolepidotus*. BGS, Kidston Collection, 3256; Fenton Coal (Langsettian); Wolley Colliery, near Barnsley, South Yorkshire, × 2.

Fig. 4. *Bothrodendron minutifolium*. NMW 92.20G10; Warren House Coal (Duckmantian); Conney Warren Opencast Pit, near Wakefield, West Yorkshire, × 1.

Fig. 5. *Lepidodendron ophiurus*. BGS, Kidston Collection 95; Sourmilk Coal (Duckmantian); New Caledonian Railway, Airdrie, Strathclyde, × 2.

## Key to Group B

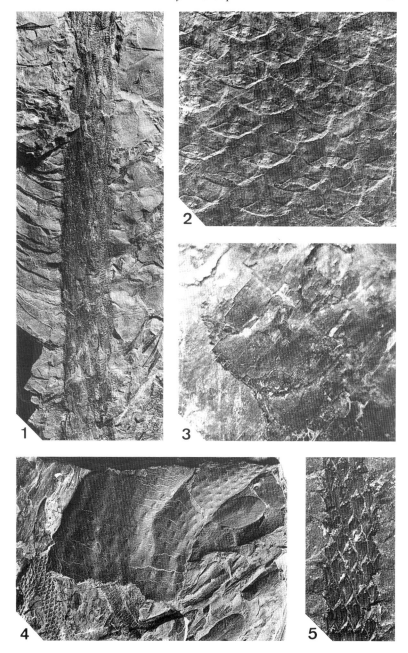

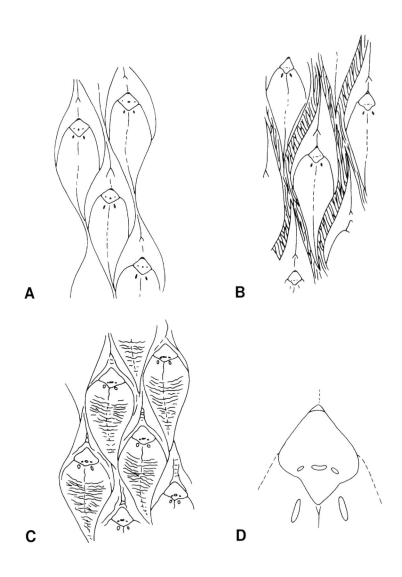

TEXT-FIG. 33. *Lepidodendron aculeatum* leaf cushions. A,B, before and after secondary expansion of the stem, × 1. C, striated cushions originally named *L. rugosa*, × 2. D, enlarged leaf scar with ligule pit aperture at upper angle and infra parichnos below. After Thomas (1970*a*).

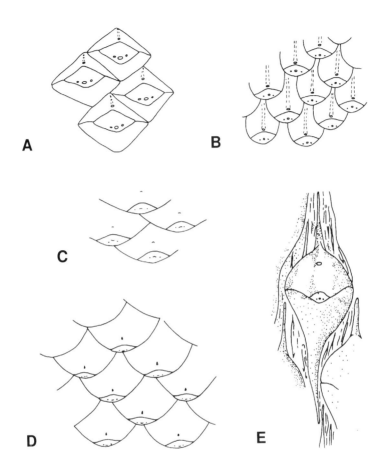

TEXT-FIG. 34. Leaf cushions of *Lepidophloios* and *Sublepidophloios*. A, B, *L. acerosus*, with undeflected and deflected leaf cushions, × 3. C, *L. laricinus*, × 1. D, *L. macrolepidotus*, × 1. E, *S. ventriculosus*, × 1. A-D after Thomas (1977), E after Hopping (1956).

23. Visible parts of leaf cushions as long, or longer, than broad ............24
    Visible parts of leaf cushion broader than long ..................................25
24. The whole visible leaf cushion downturned.....*Lepidophloios acerosus*
    (Pl. 4, fig. 3; Text-fig. 34B)
    Only the central portion of the leaf cushion downturned........................
    .......................................................... *Sublepidophloios ventricosus*
    (Text-fig. 34E)

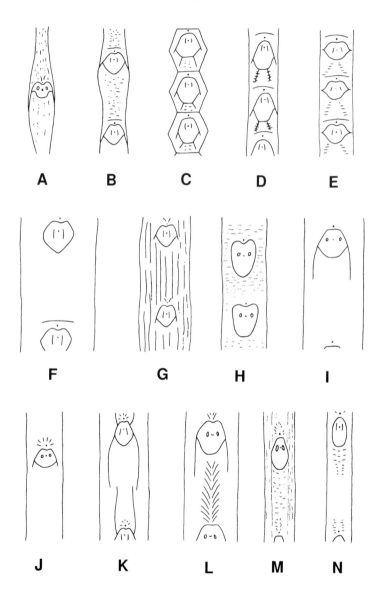

TEXT-FIG. 35. Portions of single vertical rows of *Sigillaria* leaf scars. A, *S. youngiana*. B, *S. polyploca*. C, *S. mamillaris*. D, *S. davreuxii*. E, *S. micaudii*. F, *S. cordiformis*. G, *S. scutiformis*. H, *S. cordigera*. I, *S. laevigata*. J, *S. nudicaulis*. K, *S. distans*. L, *S. schlotheimiana*. M, *S. rugosa*. N, *S. candollei*. After Crookall (1966). All × 1.

# Key to Group B

25. Leaf cushions usually < 15 mm broad............*Lepidophloios laricinus*
 (Pl. 6, fig. 2; Text-fig. 34C)
 Leaf cushions usually 20 mm or more broad ........................................
 ..............................................................*Lepidophloios macrolepidotus*
 (Pl. 6, fig. 3; Text-fig. 34D)
26. Leaf scars on hexagonal-shaped leaf cushions in vertical files..........27
 Leaf scars on vertically aligned ribs.................................................30
27. Leaf cushions about as long, or slightly longer, than broad. Lateral angles obtuse..................................................................................28
 Leaf cushions broader than long. Lateral angles acute ....................29
28. Leaf scars > 0·5 the length of the cushions. Cushion surface smooth ..
 ..........................................................................*Sigillaria elegans*
 (Text-fig. 37E)
 Leaf scars 0·5 or less the length of the cushions. Cushion surface below the scar with many short horizontal marks *Sigillaria mamillaris*
 (Text-fig. 35C)
29. Leaf scars occupying virtually the whole length of the cushion............
 ....................................................................... *Sigillaria ichthyolepis*
 (Text-fig. 37D)
 Leaf scars occupying approximately 0·5 the length of the cushion .......
 ............................................................................... *Sigillaria brardii*
 (Text-fig. 37B)
30. Leaf scars at least 1·25 times longer than broad ..............................31
 Leaf scars < 1·25 times longer than broad, or broader than long.......33
31. Lateral lines descending from the sides of the leaf scars..................33
 No lateral lines descending from the sides of the leaf scars .................
 ........................................................................… *Sigillaria candollei*
 (Text-fig. 35N)
32. Lateral lines decending about half way to the scar below. Edges of ribs smooth............................................................*Sigillaria elongata*
 (Text-fig. 36H)
 Lateral lines descending almost to the leaf scars below. Edges of ribs with short, longitudinal striations...............................*Sigillaria rugosa*
 (Pl. 7, fig. 5; Text-fig. 35M)
33. Transverse furrows visible above each leaf scar ..............................34
 No transverse furrows above the leaf scars ......................................48
34. Lateral lines extending from the sides of the leaf scars ....................35
 No lateral lines extending from the sides of the leaf scars................41
35. Edges of ribs undulating ..........................................*Sigillaria polyploca*
 (Text-fig. 35B)
 Edges of ribs straight .......................................................................36
36. Two extra basal lines extending obliquely downwards and outwards from the lower angles of hexagonal-shaped leaf scars......................37
 No extra basal lines extending downwards from the leaf scars .........39

37. Upper angle of leaf scar rounded ............................ *Sigillaria davreuxii*
    (Text-fig. 35D)
    Upper angle of leaf scar indented ....................................................... 38
38. Lateral lines short, transverse. Basal lines indistinct .............................
    ................................................................................ *Sigillaria micaudii*
    (Text-fig. 35E)
    Lateral and basal lines distinct; both directed obliquely outwards and
    downwards ........................................................ *Sigillaria reniformis*
    (Text-fig. 36D)
39. Leaf scars with rounded lateral angles. Many fine transverse and
    longitudinal marks on the middle of the ribs ............ *Sigillaria principis*
    (Text-fig. 36E)
    Leaf scars with sharp lateral edges. Many transverse/oblique marks on
    the middle of the ribs .................................................................. 40
40. Leaf scars slightly longer than broad. Lateral lines short, not
    extending much further than the base of the leaf scar. Sides of ribs
    smooth ............................................................... *Sigillaria scutellata*
    (Pl. 7, fig. 3; Text-fig. 36F)
    Leaf scars broader than long. Lateral lines extending nearly to the leaf
    scar below. Sides of ribs with longitudinal lines ..... *Sigillaria candollei*
    (Text-fig. 35N)
41. Leaf scars > 0·75 breadth of ribs ....................................................... 42
    Leaf scars < 0·75 breadth of ribs ....................................................... 45
42. Ribs ornamented ........................................................ *Sigillaria boblayi*
    (Pl. 7, fig. 1; Text-fig. 36A)
    Ribs smooth ................................................................................... 45
43. Leaf scars longer than broad ..................................... *Sigillaria latibasa*
    (Text-fig. 36G)
    Leaf scars as broad, or broader, than long .......................................... 44
44. Leaf scars less than a leaf scar length apart ............ *Sigillaria tessellata*
    (Pl. 7, fig. 4; Text-fig. 36M)
    Leaf scars more than a leaf scar length apart .... *Sigillaria transversalis*
    (Text-fig. 36J)

---

EXPLANATION OF PLATE 7

Fig. 1. *Sigillaria boblayi*. NMW 26.213G8; Two Foot Nine Coal (Duckmantian); Deep Navigation Colliery, Treharris, Mid-Glamorgan, × 1.

Fig. 2. *Sigillaria ovata*. NMW 92.20G11; Barnsley Coal (Duckmantian); Frickley Colliery, near Doncaster, South Yorkshire, × 1.

Fig. 3. *Sigillaria scutellata*. NMW 86.101G59; upper Langsettian; Cattybrook Claypit, near Almondsbury, Avon, × 1.

Fig. 4. *Sigillaria tessellata*. NMW G.1649; Coal Measures (Westphalian); Coalbrookdale, Shropshire, × 1.

Fig. 5. *Sigillaria rugosa*. BMNH V.62053; Four Foot Coal (Langsettian); St Helen's Colliery, near Workington, Cumbria, × 0·33.

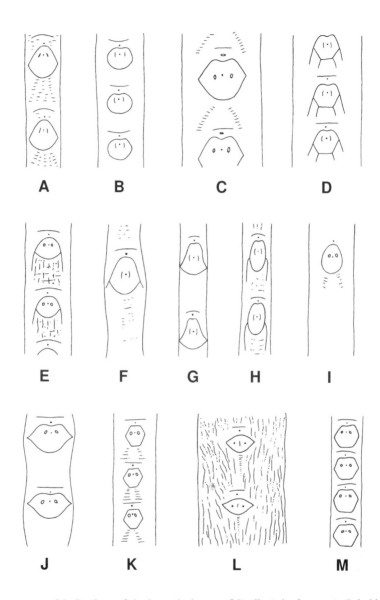

TEXT-FIG. 36. Portions of single vertical rows of *Sigillaria* leaf scars. A, *S. boblayi*. B, *S. nortonensis*. C, *S. sol*. D, *S. reniformis*. E, *S. principis*. F, *S. scutellata*. G, *S. latibasa*. H, *S. elongata*. I, *S. ovata*. J, *S. transversalis*. K, *S. sauveurii*. L, *S. kidstonii*. M, *S. tessellata*. After Crookall (1966). All × 1.

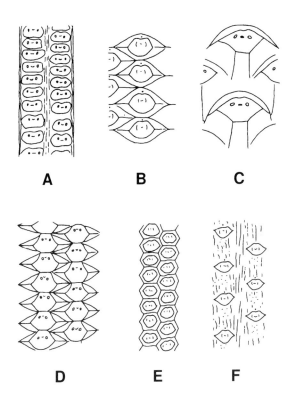

TEXT-FIG. 37. Portions of vertical rows of *Sigillaria* leaf scars. A, *S. lorwayana*. B, *S. brardii*. C, *S. macmurtriei*. D, *S. ichthyolepis*. E, *S. elegans*. F, *S. reticulata*. After Crookall (1966). All × 1.

45. Leaf scars with rounded upper angles ................................................46
    Leaf scars with indented upper angles ................................................47
46. Leaf scars transversely oval, lateral edges acute. Rib surface covered with longitudinal lines ...............................................*Sigillaria kidstonii*
    (Text-fig. 36L)
    Leaf scars hexagonal. Basal lines extending obliquely downwards from the lower angles of the leaf scars ....................*Sigillaria sauveurii*
    (Text-fig. 36K)
47. Leaf scars with rounded lateral and basal angles. Rib surface smooth . ........................................................................ *Sigillaria nortonensis*
    (Pl. 8, fig. 1; Text-fig. 36B)

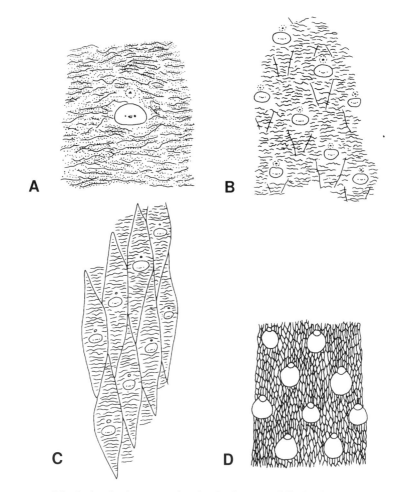

TEXT-FIG. 38. *Bothrodendron* stem showing leaf sears and ligule pit apertures. A, *B. minutifolium*, surface of large stem. B, C, *B. minutifolium*, small branches. D, *B. punctatum*, surface of large stem. After Thomas (1967*b*). All × 5.

EXPLANATION OF PLATE 8

Fig. 1. *Sigillaria nortonensis*. NMW 90.8G2; Farrington Formation (upper Westphalian D); Lower Writhlington Colliery Tip, Radstock, Avon, × 1.

Fig. 2. *Sigillaria lorwayana*. NMW 90.9G3; Farrington Formation (upper Westphalian D); Kilmersdon Colliery Tip, Radstock, Avon, × 1.

1

2

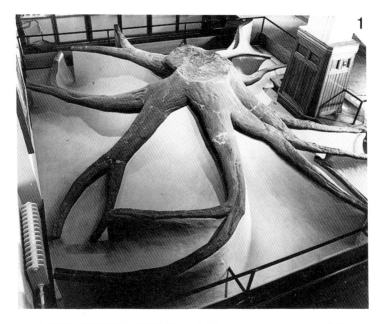

Leaf scars cordate. Rib surface smooth ............... *Sigillaria cordiformis*
(Text-fig. 35F)
Leaf scars with obtuse-angled lateral angles and rounded lower angle; basal lines extending obliquely downwards from the leaf scars ............
..................................................................................*Sigillaria sol*
(Text-fig. 36C)
48. Lateral lines extending from the sides of the leaf scar ...................49
    No lateral lines extending from the sides of the leaf scar ................54
49. Lateral lines not extending below the base of the leaf scar .............50
    Lateral lines extending below the base of the leaf scar ..................51

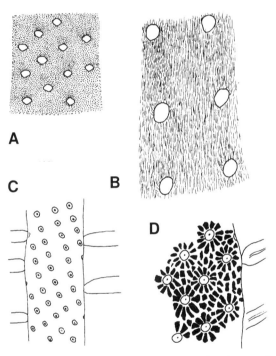

TEXT-FIG. 39. *Cyclostigma* stems surfaces and *Stigmaria* rhizophore surfaces. A, *C. cambricum*, × 2. B, *C. macconochiei*, × 2. C, *S. ficoides*, × 1. D, *S. stellata*, × 1. After Crookall (1966).

EXPLANATION OF PLATE 9

Fig. 1. *Stigmaria* sp., Manchester Museum, photo. J. Watson.
Fig. 2. *Stigmaria ficoides*. NMW, David Davies Collection 22.113G97; Pentre Coal (Bolsovian); Gilfach Goch, Mid-Glamorgan, × 0·5.

50. Rib surface with many longitudinal lines ............*Sigillaria scutiformis*
 (Text-fig. 35G)
 Rib surface smooth ................................................*Sigillaria nudicaulis*
 (Text-fig. 35J)
51. Rib surface ornamented ................................................................52
 Rib surface smooth ........................................................................53
52. Ribs ornamented with small longitudinal lines. Rib margins undulate
 ...........................................................................*Sigillaria youngiana*
 (Text-fig. 35A)
 Ribs ornamented centrally with 2 rows of oblique, inwardly directed, lines. Rib margins straight............................*Sigillaria schlotheimiana*
 (Text-fig. 35L)

A  B

TEXT-FIG. 40. A, halonial scar; NMW 68.163G1; Coal Measures (Westphalian); United Kingdom (further details not available). B, *Artisia*; NMW 40.315G2; Pentre Coal (lower Bolsovian); Coedely Colliery, Tonyrefail, Mid-Glamorgan. Both × 1.

EXPLANATION OF PLATE 10

*Ulodendron majus*. NMW 86.59G94; Coal Measures (Westphalian); Uphall, Edinburgh, × 1.

## Key to Group B

53. Extra lines extending upwards from the apices of the scars to within those extending downwards from the scars above......*Sigillaria distans* (Text-fig. 35K)
    No extra lines extending upwards from the scar.....*Sigillaria laevigata* (Text-fig. 35I)
54. Leaf scars 0·75 breadth of the ribs.......................*Sigillaria lorwayana* (Pl. 8, fig. 2; Text-fig. 37A)
    Leaf scars not as broad as the ribs................................................55

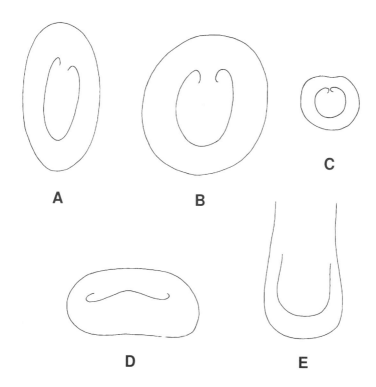

TEXT-FIG. 41. *Caulopteris* and *Megaphyton* leaf scars. A, *C. anglica*. B, *C. cyclostigma*. C, *C. arberi*. D, *M. gwynnevaughanii*. E, *M. frondosum*. After Crookall (1955). All × 1.

---

EXPLANATION OF PLATE 11

*Caulopteris anglica*. BMNH V.3088; Farrington or Radstock formations (upper Westphalian D); Radstock, Avon, × 1.

55. Leaf scars oval ..................................................Sigillaria ovata
       (Pl. 7, fig. 2; Text-fig. 36I)
    Leaf scars cordate ..............................................Sigillaria cordigera
       (Text-fig. 35H)
56. Ligule pits present above leaf scars........................................57
    Ligule pits absent.................................................................59
57. Leaf scars with acute lateral angles ....................Sigillaria reticulata
       (Text-fig. 37F)
    Leaf scars with rounded sides............................................58
58. Bark ornamented with irregular, horizontal, undulating wrinkles
    ................................................................... Bothrodendron minutifolium
       (Pl. 6, fig. 4; Text-fig. 38A-C)

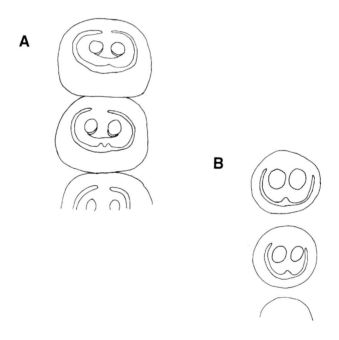

TEXT-FIG. 42. Vertical rows of *Artisophyton* leaf scars. A, *A. goldenbergii*. B, *A. approximatum*. After Crookall (1955). All × 0·5.

     Bark ornamented with irregular, vertical wrinkles....................................
     .................................................................. *Bothrodendron punctatum*
     (Text-fig. 38D)
 59. Bark smooth with fine punctae........................*Cyclostigma cambricum*
     (Text-fig. 39A)
     Bark ornamented with longitudinal striae ...*Cyclostigma macconochiei*
     (Text-fig. 39B)
 60. Scars < 10 mm in diameter and unassociated with scars of any other shape..............................................................................................................61
     Scars > 10 mm in diameter and associated with smaller rhomboidal or rounded scars..............................................................................................62
 61. Scars on axis with little surface ornamentation.........*Stigmaria ficoides*
     (Pl. 9; Text-fig. 39C)
     Scars surrounded by radiating lines..........................*Stigmaria stellata*
     (Text-fig. 39D)
 62. Cup-shaped depressions on stems, with a central or excentric slightly projecting boss .........................................................ulodendroid scar
     (Pl. 10)
     Prominent tubercle sometimes with a central depression...halonial scar
     (Text-fig. 40A)
 63. Stems with more than two rows of longitudinal rows of leaf scars, forming spirals or whorls around the stem.........................................64
     Stems with only two rows of longitudinal rows of leaf scars, one on either side of the stem..........................................................................66
 64. Leaf scars oval, averaging 60 mm long and 40 mm wide.....................
     ..................................................................... *Caulopteris anglica*
     (Pl. 11; Text-fig. 41A)
     Leaf scars more or less circular..........................................................65
 65. Leaf scars > 40 mm in diameter. Vascular scars 'vase'-shaped, with relatively wide neck 6–10 mm wide................*Caulopteris cyclostigma*
     (Text-fig. 41B)
     Leaf scars < 40 mm in diameter. Vascular scars 'vase'-shaped, with narrow neck 1–3 mm wide........................................*Caulopteris arberi*
     (Text-fig. 41C)
 66. Single 'vase'- or W-shaped vascular scar..........................................67
     Two or more vascular scars. Outer scar(s) form(s) a closed trace with a deep indentation on the lower side; or consist(s) of two closed traces. Smaller S-shaped scars occur within outer scar.....................68
 67. Leaf scars longer than wide, up to 100 mm long ...................................
     ..................................................................... *Megaphyton frondosum*
     (Text-fig. 41E)
     Leaf scars wider than long, < 60 mm long ..........................................
     ..............................................................*Megaphyton gwynnevaughanii*
     (Text-fig. 41D)

68. Oval to subrectangular leaf scars, > 35 mm long and > 45 mm wide.
Adjacent scars may be contiguous ............... *Artisophyton goldenbergii*
(Text-fig. 42A)
Oval to circular leaf scars, < 35 mm long and < 45 mm wide. Adjacent
scars never contiguous ........................... *Artisophyton approximatum*
(Text-fig. 42B)

# KEY TO GROUP C:
# LEAVES BORNE ON STEMS IN WHORLS

Except for the lycopsid sporophylls *Lepidostrobophyllum* at node 2, this group consists of sphenopsid foliage. Broadly, two types can be recognized.

Firstly, there are calamite leaves, which are elongate, linear or lanceolate, with a single longitudinal vein. They belong to the form-genera *Annularia* or *Asterophyllites*, and are classified in the Order Equisetales. Characters used to separate species of these leaves include their length, the shape of their apex, where along their length is their widest part, and how many occur in each whorl. In some of the smaller-leaved species, the angle of attachment of the leaves to the stem can be important, although this feature is not always visible in the fossils.

The second type consists of broader, cuneate or ovate leaves with a dichotomizing vein system. All of the British species of this type belong to the form-genus *Sphenophyllum*, and are normally placed in their own, extinct order, the Bowmanitales. Species may be recognized on the length of the leaves and their length : breadth ratio. Also of importance is the degree to which their distal margin is incised, and the shape of the distal lobes/teeth, although these features can show considerable variation in different parts of the same plant (Batenburg 1977).

There are numerous records of this type of foliage from the British Coal Measures. However, the only attempt at a monographic treatment in recent years was by Crookall (1969), and this is incomplete and poorly illustrated. Rather more complete accounts are available for some of the foreign fossil floras and these provide a better introduction to the systematics and biology of the group (e.g. Abbott 1958; Storch 1966; Batenburg 1977, 1981).

1. Linear to lanceolate leaves with a single longitudinal vein .................. 2
   Cuneate to ovate leaves with dichotomous, radiating veins ............... 12
2. Leaves > 5 mm wide at widest point .................... *Lepidostrobophyllum*
   (see Group G)
   Leaves < 5 mm wide at widest point ................................................. 3
3. More or less linear leaves which are widest (if anywhere) in the proximal one third ............................................................................ 4
   Linear-lanceolate or spatulate leaves which are widest in the middle or distal part ........................................................................................ 8

# Key to Group C

4. More than 25 leaves in a whorl. Each leaf > 50 mm long
.................................................... *Asterophyllites longifolius*
(Pl. 13, fig. 3; Text-fig. 43D)
Less than 25 leaves in a whorl. Each leaf < 50 mm long ...................5
5. More than 10 leaves in a whorl. Each leaf usually > 10 mm long
.................................................... *Asterophyllites equisetiformis*
(Pl. 13, fig. 2; Text-fig. 43A)
Less than 10 leaves in a whorl. Each leaf < 10 mm long ...................6

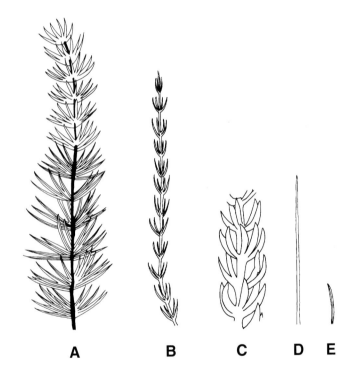

TEXT-FIG. 43. *Asterophyllites* shoots and leaves. A, *A. equisetiformis*, × 1. B, *A. lycopodioides*, × 1. C, *A. charaeformis*, × 10. D, *A. longifolius*, × 4. E, *A. grandis*, × 1. A, C-E after Abbott (1958), B after Crookall (1969).

EXPLANATION OF PLATE 12

Figs 1, 2. Pteridosperm axes. 1, NMW 87.20G57, ×1. 2, NMW 87.20G59, × 0·5.
Fig. 3. *Sphenophyllum* axis. NMW 87.20G137, × 1.
Fig. 4. *Sigillariostrobus* axis. NMW 87.20G4, × 1.
All specimens from upper Duckmantian, Howgill Head Quarry, Whitehaven, Cumbria.

6. More than 6 leaves in a whorl. Each leaf > 3 mm long ........................
   ............................................................. *Asterophyllites grandis*
   (Text-fig. 43E)
   Less than 6 leaves in a whorl. Each leaf < 3 mm long ........................ 7
7. Leaves extend out at right angles from stem and are strongly arched
   .................................................................*Asterophyllites charaeformis*
   (Text-fig. 43C)
   Leaves attached obliquely to stem and are not strongly arched ............
   ............................................................. *Asterophyllites lycopodioides*
   (Text-fig. 43B)

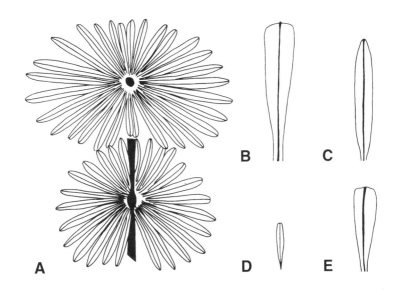

TEXT-FIG. 44. *Annularia* shoots and leaves. A, *A. stellata*, × 3. B, *A. mucronata*, × 1. C, *A. radiata*, × 1. D, *A. galioides*, × 1. E, *A. sphenophylloides*, × 1. After Abbott (1958).

EXPLANATION OF PLATE 13

Fig. 1. *Sphenophyllum cuneifolium*. BMNH V.63667; Middle Coal Measures (middle Duckmantian); Rhigos, near Hirwaun, Mid-Glamorgan, × 1.
Fig. 2. *Asterophyllites equisetiformis*. NMW 90.8G3; Farrington Formation (upper Westphalian D); Lower Writhlington Colliery Tip, Radstock, Avon, × 1.
Fig. 3. *Asterophyllites longifolius*. BMNH V.1345; Ten Foot Ironstone (Duckmantian); Coseley, Dudley, West Midlands, × 1.

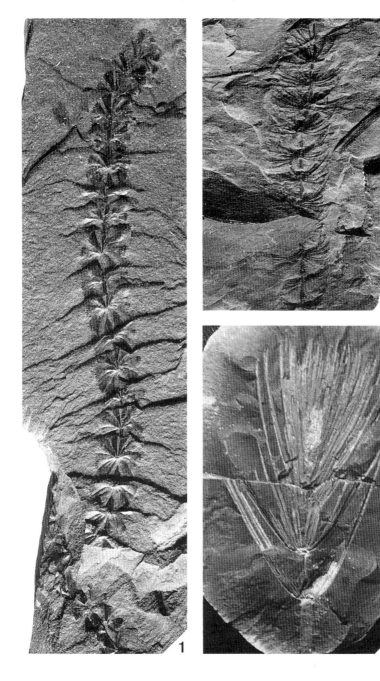

8. Leaves widest in middle .................................................................. 9
   Leaves widest above middle ............................................................ 10
9. Leaves 6–15 mm long, with acuminate apices .......... *Annularia radiata*
   (Text-fig. 44C)
   Leaves 1–4 mm long, with blunt apices ................ *Annularia galioides*
   (Text-fig. 44D)
10. Oblanceolate leaves, up to 75 mm long .................... *Annularia stellata*
    (Pl. 14; Text-fig. 44A)
    Spatulate leaves, never more than 25 mm long ................................. 11
11. Leaves may be more than 12 mm long. Some lateral leaves in whorl more elongate than others ........................................ *Annularia mucronata*
    (Text-fig. 44B)
    Leaves never more than 12 mm long. Leaves in whorl of equal length ........................................................ *Annularia sphenophylloides*
    (Text-fig. 44E)

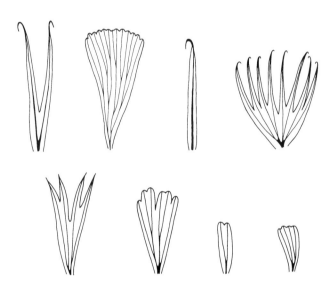

TEXT-FIG. 45. *Sphenophylum emarginatum*, showing variation in leaf shape. After Batenburg (1977). All × 3.

EXPLANATION OF PLATE 14

*Annularia stellata*. NMW 90.8G4; Farrington Formation (upper Westphalian D); Lower Writhlington Colliery Tip, Radstock, Avon, × 1.

# Key to Group C

12. Whorls consist of leaves of markedly unequal length..........................
    ................................................Sphenophyllum oblongifolium
    (Text-fig. 46D)
    Whorl consists of leaves of more or less equal length......................13
13. Length : breadth ratio of leaves about 2 ...........................................14
    Length : breadth ratio of leaves 3 or more ........................................18
14. Lateral margins of leaves more or less concave................................15
    Lateral margins of leaves straight to convex .....................................16

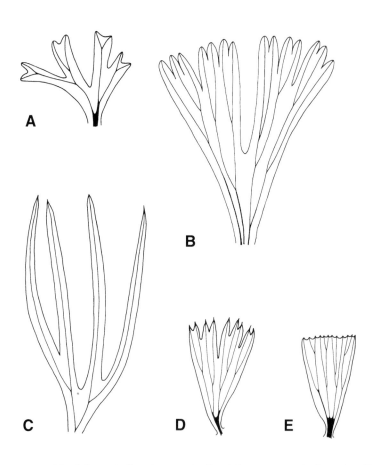

TEXT-FIG. 46. Sphenophyllum leaves. A, S. trichomatosum. B, S. majus. C, S. myriophyllum. D, S. oblongifolium. E, S. cuneifolium. After Abbott (1958). All × 3.

Key to Group C 97

15. Leaves 8–10 mm long. Shallow, bluntly acuminate teeth on distal margin. Also, a median cleft on distal margin, extending for < 0·33 of leaf length..................................................*Sphenophyllum emarginatum*
(Text-fig. 45)
Leaves about 5 mm long. Distal margins of leaves deeply incised, with a median cleft extending for > 0·5 of leaf length..........................
..........................................................*Sphenophyllum trichomatosum*
(Text-fig. 46A)
16. Leaves with bluntly acuminate teeth on distal margin.........................
..................................................................*Sphenophyllum emarginatum*
(Text-fig. 45)
Leaves with sharply pointed teeth on distal margin.........................17
17. Leaves < 12 mm long.........................*Sphenophyllum cuneifolium*
(Pl. 13, fig. 1; Text-fig. 46E)
Leaves > 12 mm long............................................*Sphenophyllum majus*
(Text-fig. 46B)
18. Distal margins of leaves deeply lobed, each lobe being of more or less equal length...............................................*Sphenophyllum trichomatosum*
(Text-fig. 46A)
Distal margins of leaves deeply lobed, with lobes of unequal length ..
....................................................................................................................19
19. Clefts between lobes on distal margin extend for two-thirds or more of length of leaf. Lobes very slender and widely spreading......................
..................................................*Sphenophyllum myriophyllum*
(Text-fig. 46C)
Clefts between lobes on distal margin extend for < 0·67 of length of leaves. Lobes rather broad and not widely spreading ....................20
20. Leaves with sharply pointed lobes..............*Sphenophyllum cuneifolium*
(Text-fig. 46E)
Leaves with bluntly acuminate lobes........*Sphenophyllum emarginatum*
(Text-fig. 45)

# KEY TO GROUP D: BRANCHING STEMS COVERED WITH SMALL TAPERING OR STRAP-LIKE 'LEAVES', BELONGING TO THE LYCOPSIDS AND CONIFERS

Except for one conifer (*Walchia*), these are leafy shoots of the lycopsids. Particularly in *Lepidodendron* and *Sigillaria*, the larger stems are not normally preserved with the leaves still attached; such stems showing leaf scars key out in Group B. Exceptions can sometimes occur in these form-genera (e.g. Leary and Thomas 1989), while other form-genera such as *Ulodendron* appear never to have shed their leaves (Thomas 1967*a*). Among the arborescent lycopsids most commonly represented in the Coal Measures adpression record, however, leaves are only found attached to terminal shoots.

Such terminal shoots of the arborescent lycopsids tend to have a superficially similar appearance. They are, therefore, impossible to interpret as meaningful species and some workers (e.g. Crookall 1964) have taken the extreme view that they should all be grouped together as one species, i.e. *L. ophiurus*. Although that view is not subscribed to here, it must be stressed that the names are artificial, and must not be thought of as representing true biological species.

The herbaceous lycopsids, *Selaginellites* and *Lycopodites*, are very rare and can be distinguished on their leaf arrangements. *Lycopodites* can be readily confused with the very slender terminal shoots of *Lepidodendron* and *Bothrodendron*.

Conifer remains are rare in the British Coal Measures, and tend to be restricted to red-beds, such as the Keele Formation. There has been some debate recently about the nomenclature of such conifer fragments

---

EXPLANATION OF PLATE 15

Figs 1, 2. *Walchia* sp. BMNH V.5965; Westphalian D(?); Granville Pit, Donnington, Shropshire, × 3.

Fig. 3. *Selaginellites gutbieri*. NMW 90.8G7; Farrington Formation (upper Westphalian D); Lower Writhlington Colliery Tip, Radstock, Avon, × 1.

Fig. 4. *Lepidodendron lycopodioides*. NMW G.1629; roof of Brithdir Seam (lower Westphalian D); Cefn Brithdir Pit, Rhymney, Mid-Glamorgan, × 1.

Fig. 5. *Calamostachys ramosa*. NMW 87.20G10; upper Duckmantian; Howgill Head Quarry, Whitehaven, Cumbria, × 2.

Key to Group D 99

(reviewed by Mapes and Rothwell 1991). In this volume we have followed Mapes and Rothwell, and used the form-genus *Walchia* for fragments of conifer shoot that show neither evidence of cuticles or fructifications, while recognizing that the names may have to be changed in the future.

1. Axes with elongate, strap-like structures attached ............................. 2
   Axes with small, tapering leaves attached .......................................... 3
2. Strap-like structures attached to axes with small, rhomboidal marks ...
   ................................................................................................... lycopsid axes
   Strap-like structures attached to axes with small, rounded marks with a single central mark ........................................................... *Stigmaria*
   (see Group B)

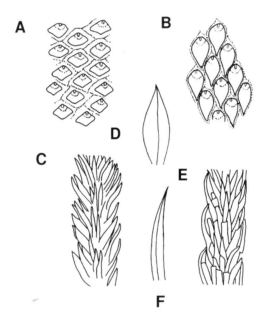

TEXT-FIG. 47. Leafy lycophyte shoots. A, *Ulodendron majus*, showing leaf bases with ligule pits in their upper angles, × 1. B, *U. landsburgii*, showing leaf bases with ligule pits in their upper angles, × 1. C,D, *Bothrodendron minutifolium*, shoot × 2, leaf × 4. E,F, *Lepidodendron* sp., the leafy shoot of *Flemingites olryi*, shoot × 2, leaf × 4. A after Thomas (1967*a*), B after Thomas (1968), C-F after Thomas (1967*b*).

3. Leafy shoots with lateral branches .................................*Walchia* sp.
   (Pl. 15, figs 1–2)
   Leafy shoots dichotomizing or unbranched...........................4
4. Leaves of two sizes, shoots with adjacent ranks of each leaf type ........
   ........................................................*Selaginellites gutbieri*
   (Pl. 15, fig. 3)
   Leaves similar in appearance, arranged spirally on shoots ..................5
5. Stem 1 mm or less in diameter, leaves 2–3 mm long ...........................
   ..........................................................*Lycopodites pendulus*
   Stem much more than 1 mm in diameter, leaves longer than 5 mm....6

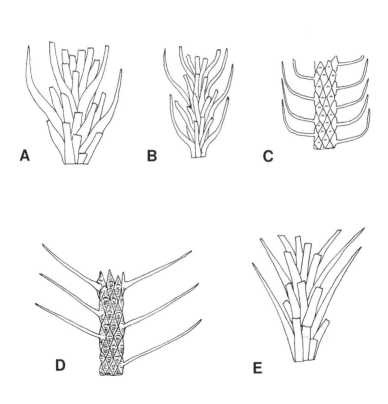

TEXT-FIG. 48. *Lepidodendron* leafy shoots. A, *L. acutum*. B, *L. simile*. C, *L. ophiurus*. D, *L. worthenii*. E, *L. lycopodioides*. All × 1.

TEXT-FIG. 49. *Lepidodendron acutum*; David Davies Collection, NMW 22.114G128; Pentre Coal (lower Bolsovian); Gilfach Goch, Mid-Glamorgan, × 1.

6. Leaves attached to the upper angle of broad, diamond-shaped leaf cushions .................................................................................... 7
   Leaves attached in the upper half of fusiform leaf cushions ............... 8
7. Leaf cushions longer than broad ..................... *Ulodendron landsburgii*
   (Text-fig. 47B)
   Leaf cushions roughly isodiametric ......................... *Ulodendron majus*
   (Pl. 10; Text-fig. 47A)
8. Leaves depart at > 45° to stem axis ..................................................... 9
   Leaves depart at < 45° to stem axis ................................................... 10
9. Leaves more or less straight ......................... *Lepidodendron worthenii*
   (Text-fig. 48D)
   Leaves depart at about right-angles then bend upwards (sickle-shaped)
   ...................................................................... *Lepidodendron ophiurus*
   (Pl. 6, fig. 5; Text-fig. 48C)
10. Leaves distinctly S-shaped .............................................................. 11
    Leaves curved but not distinctly S-shaped ...................................... 12
11. Leaves broader than 3 mm at base ................... *Lepidodendron acutum*
    (Text-figs 48A, 49)
    Leaves narrower than 3 mm at base .................... *Lepidodendron simile*
    (Text-fig. 48B)
12. Leaves broader than 3 mm at base ......... *Lepidodendron lycopodioides*
    (Pl. 4, fig. 4; Pl. 15, fig. 4; Text-fig. 48E)
    Leaves narrower than 3 mm at base ................................................ 13
13. Leaves curved inwards towards the stem ..... *Bothrodendron punctatum*
    (Text-fig. 38D)
    Leaves curved outwards from the stem .... *Bothrodendron minutifolium*
    (Text-fig. 47C,D)

# KEY TO GROUP E: PARTS OF FERN-LIKE FRONDS

Some of the fronds included in this group undoubtedly belong to true ferns, and have pteridophytic sporangia attached to the pinnules, but others belong to progymnosperms and gymnosperms (the biology of these groups is discussed by Thomas and Spicer 1987). The affinities of the principal form-genera are as follows.

Ferns: *Bertrandia, Corynepteris, Crossotheca, Cyathocarpus, Hymenophyllites, Lobatopteris, Oligocarpia, Pecopteris, Polymorphopteris, Renaultia, Sphenopteris (pro parte), Sphyropteris, Urnatopteris, Zeilleria.*

Progymnosperms: *Noeggerathia.*

Pteridosperms: *Alethopteris, Callipteridium, Cyclopteris, Dicksonites, Eusphenopteris, Fortopteris, Karinopteris, Laveineopteris, Linopteris, Lonchopteris, Lyginopteris, Macroneuropteris, Margaritopteris, Mariopteris, Neuralethopteris, Neuropteris, Odontopteris, Palmatopteris, Paripteris, Reticulopteris.*

The affinities of some members of the group, however, are still unclear, such as *Desmopteris* and some species of *Sphenopteris.*

Many of these genera, such as *Neuropteris* and *Sphenopteris,* were originally defined purely on the basis of pinnule shape and venation (e.g. Brongniart 1828–1838) and so could be readily recognized in fragments of the type frequently found in the Coal Measures. As such, they were clearly artificial concepts and were soon recognized to be unsatisfactory. It would be normal to seek evidence from reproductive structures to make the taxonomy of these fronds more natural but, except in the ferns (e.g. Brousmiche 1983), they are rarely found attached. Instead, work on frond architecture (i.e. the pattern of branching of the rachises within the frond) and epidermal structure has helped establish a more natural classification. Examples of this include the separation of *Callipteridium* from *Alethopteris* (e.g. Wagner 1965) and *Karinopteris* from *Mariopteris* (Boersma 1972) based on frond architecture, and the use of a combination of frond architecture and cuticles to separate *Laveineopteris,*

---

EXPLANATION OF PLATE 16

*Neuropteris heterophylla.* BMNH V.1867; Duckmantian; Clay Cross, Derbyshire, ×1.

*Neuralethopteris, Paripteris* and *Macroneuropteris* from *Neuropteris* (Laveine 1967; Cleal *et al.* 1989). The use of such characters clearly requires exceptional preservation, either of large specimens for frond architecture (e.g. Zodrow and Cleal 1988; Cleal and Shute 1991) or low rank of coalification for cuticles (Barthel 1961, 1962; Cleal and Zodrow 1989; Cleal and Shute 1991, 1992; Zodrow and Cleal 1993). In the vast majority of the fossils found in the Coal Measures such characters cannot be observed.

Most fossils must therefore be first assigned to a particular species and only then will its generic position be known. Important characters for distinguishing the species, and used in the key below, are the form of the pinnules (including their shape, the degree of fusion to the rachis, and to what extent they are lobed), the strength of the midvein, details of the lateral veins (including their density along and angle to the pinnule margin, the number of times they fork, and whether or not they are anastomosed), and the shape of the pinnae. If present, the form and number of the sporangia can also be important.

Many of the species show a considerable degree of morphological variation, and there is often some overlap between species. It has been impossible to take all of this variation into account in the key and so, if it is used to identify isolated, small fragments, it may produce misleading results. More robust results will be obtained by trying to combine the specimens from a particular locality and horizon into what appear to be 'natural' groups, and then trying to key out the more typical forms. Establishing these 'natural' groups is, of course, not easy for anyone unfamiliar with the Coal Measures floras, but experience will eventually improve the results. For anyone intending to study these fossils in any detail, it is advisable for them to investigate the appropriate literature, to get some idea as to how other palaeobotanists have grouped specimens into particular species.

The best monographs published in Britain are by Kidston (1923–1925) and Crookall (1955, 1957). They are useful records of the British floras, but use systematics which are now seriously out of date and have to be treated with a great deal of caution. To get a better idea of how Coal Measures plant fossils are classified today, it is better to consult the foreign literature, such as Patteisky (1957); Dalinval (1960); Buisine (1961); Laveine (1967); Laveine *et al.* (1977); Wagner (1968); Boersma (1972); van Amerom (1975) and Brousmiche (1983). These are all extensively illustrated and provide excellent introductions to the systematics of most of the form-genera covered by the following key.

1. Pinnules mostly with unlobed or only shallowly lobed margins ........2
   Pinnules mostly with deeply lobed margins .................................57

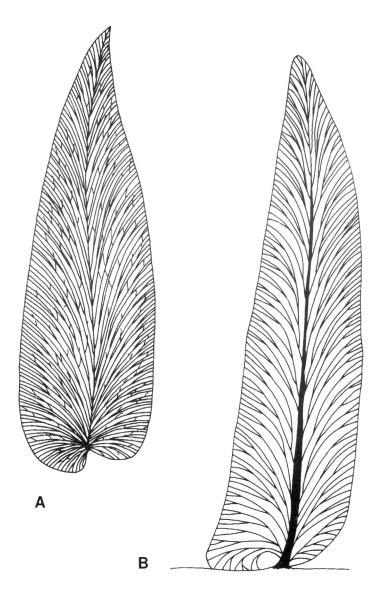

TEXT-FIG. 50. *Macroneuropteris* pinnules. A, *M. scheuchzeri*. B, *M. macrophylla*. After Crookall (1957). Both × 3.

2. Pinnules attached to rachis by > 0·5 their width ................................. 3
   Pinnules isolated, or attached to rachis by < 0·5 their width ............... 5
3. Pinnules, when unlobed, usually > 10 mm long; occasionally pinnules 5–10 mm long, but then are squat with a length : breadth ratio about 2 ............................................................................................ 4
   Pinnules, when unlobed, usually < 5 mm long; occasionally pinnules 5–10 mm long, but then are slender with a length : breadth ratio >1 : 2 ........................................................................................................... 117
4. Veins meet pinnule margin at 60–90°. When pinnules are lobed low in pinna, basiscopic and acroscopic lobes more or less equally developed ............................................................................................. 38
   Veins meet pinnule margin at < 60°. When pinnules are lobed low in pinna, basiscopic lobe is more prominently developed ................... 84

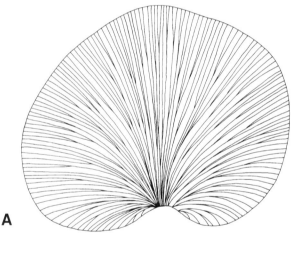

TEXT-FIG. 51. *Cyclopteris* leaflets. A, *C. orbicularis*. B, *C. fimbriata*. A after Crookall (1957), B after Cleal (1985). Both × 1.

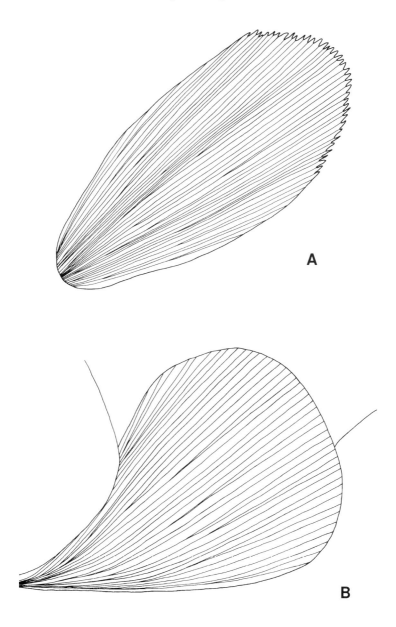

TEXT-FIG. 52. *Noeggerathia* leaves. A, *N. chalardii*. B, *N. foliosa*. After Boureau (1964). Both × 3.

5. Isolated pinnules ..................................................................6
   Pinnules attached to a rachis ..............................................11
6. Circular to reniform pinnules with no midvein .....................7
   More elongate pinnules with midvein ..................................10
7. Pinnules < 10 mm long ........................................................8
   Pinnules > 10 mm long ........................................................9
8. Hairs visible on pinnule surface ..........*Macroneuropteris scheuchzeri*
   (Pl. 18, fig. 2; Text-fig. 50A)
   No hairs visible on pinnules surface .............................*Paripteris* sp.
9. Pinnule with smooth margin .........................*Cyclopteris orbicularis*
   (Pl. 17, fig. 1; Text-fig. 51A)
   Pinnule with fimbriate or laciniate margin ........*Cyclopteris fimbriata*
   (Text-fig. 51B)
10. Large linguaeform pinnules with numerous hairs on surface ..............
    ......................................................*Macroneuropteris scheuchzeri*
    (Pl. 18, fig. 2; Text-fig. 50A)
    Pinnules showing no hairs on surface .................................34
11. Pinnules lying at *c*. 45° to rachis. Veins radiating from point of attachment at base of pinnules, and without a midvein ..................12
    Pinnules lying at 60–90° to rachis. Nervation consists of midvein with lateral veins ..................................................................13
12. Pinnules with deeply dentate distal margin .........*Noeggerathia foliosa*
    (Text-fig. 52B)
    Pinnules with shallowly dentate distal margin*Noeggerathia chalardii*
    (Text-fig. 52A)
13. Pinnules narrowly attached to rachis low in pinna, becoming more fused to rachis towards pinna terminal; pinna terminal with single apical pinnule; rachis longitudinally striate ...................................14

EXPLANATION OF PLATE 17

Fig. 1. *Cyclopteris orbicularis*. NMW 90.8G24; Farrington Formation (upper Westphalian D); Lower Writhlington Colliery Tip, Radstock, Avon, × 1.

Fig. 2. *Corynepteris angustissima*. NMW 86.101G31; upper Langsettian; Cattybrook Claypit, near Almondsbury, Avon, × 4.

Fig. 3. *Neuropteris dussartii*. NMW, David Davies Collection 9857; No. 1 Rhondda Coal (Bolsovian); Llwyn-y-pia Level, Glamorgan Collieries, near Pontypridd, Mid-Glamorgan, × 1.

Fig. 4. *Laveineopteris rarinervis*. NMW, David Davies Collection 8089; No. 3 Llantwit Coal (Westphalian D); Rhondda Valley, near Pontypridd, Mid-Glamorgan, × 1.

Fig. 5. *Laveineopteris tenuifolia*. NMW, David Davies Collection 10957; Three Coal Seam (Bolsovian); Glynogwr Colliery, near Pontypridd, Mid-Glamorgan, × 1.

# Key to Group E

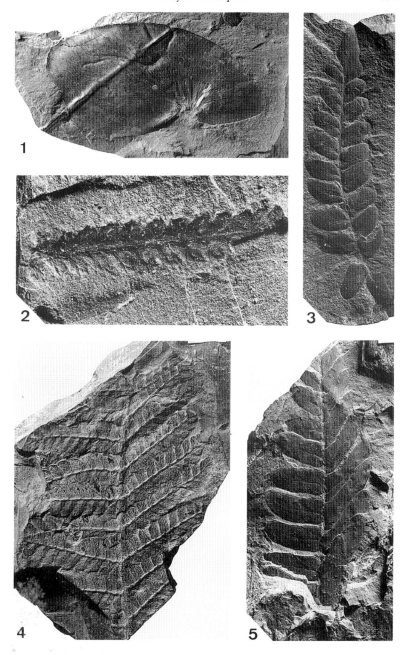

Pinnules narrowly attached to rachis throughout pinna; pinna terminal with a pair of apical pinnules; rachis with numerous punctae ...............................................................................................................33

14. Midvein extending for almost entire length of pinnule ...................15
    Midvein extending for 0·75 or less of pinnule length......................18
15. Lateral veins meet pinnule margin at < 60°. Nervation density < 30 veins per cm. Pinnules attached to rachis at 60° or less, and asymmetrical about midvein...................................*Mariopteris sauveurii*
    (Text-fig. 71D)
    Lateral veins meet pinnule margin at 60–90°. Nervation density > 30 veins per cm. Pinnules attached to rachis at 60–90°, and more symmetrical about midvein ................................................................16
16. Pinnules generally < 10 mm long; fairly irregular nervation, with vein density generally < 50 veins per cm.*Neuralethopteris schlehanii*
    (Text-fig. 53E)
    Pinnules generally > 10 mm long; regular, dense nervation, with vein density generally > 50 veins per cm ...................................................17
17. Lateral veins broadly arched between midvein and pinnule margin; pinnules generally oblong with a rounded apex ...................................
    .................................................................*Neuralethopteris jongmansii*
    (Text-fig. 53F)
    Lateral veins arch mainly near midvein, and then extend in a straight line towards pinnule margin; pinnules generally lanceolate with an acuminate or bluntly acuminate apex ......*Neuralethopteris rectinervis*
    (Text-fig. 53G)
18. Pinnules generally with constricted base, except high in the pinna .19
    Pinnules showing a much greater tendency to become fused to the rachis............................................................................................28
19. Lateral veins rather flexuous, often widely forking.........................20
    Lateral veins broadly arched and narrowly forking.........................21
20. Small pinnules, generally < 13 mm long; very widely forking veins producing vein density < 35 per cm.............*Laveineopteris rarinervis*
    (Pl. 17, fig. 4; Text-fig. 53B)
    Pinnules generally > 13 mm long with vein density > 35 per cm ...26
21. Vein density generally < 35 veins per cm........................................22
    Vein density generally > 35 veins per cm........................................23
22. Average pinnule length ⩽ 10 mm....................................................23
    Average pinnule length > 10 mm.....................................................24
23. Pinnules rather squat and subrectangular............*Neuropteris dussartii*
    (Pl. 17, fig. 3; Text-fig. 54D)
    Pinnules rather rounded, or subfalcate to linguaeform in the larger forms ..............................................................*Laveineopteris loshii*
    (Text-fig. 53C,D)

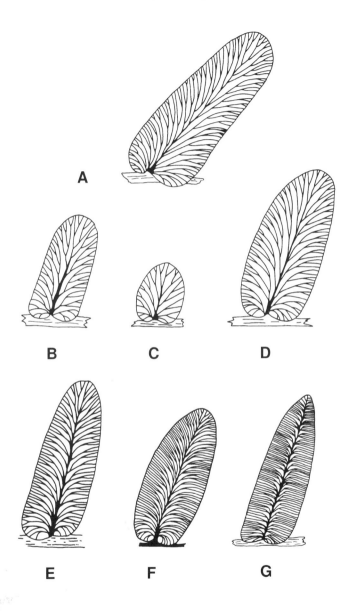

TEXT-FIG. 53. *Laveinopteris* and *Neuralethopteris* pinnules. A, *L. tenuifolia*. B, *L. rarinervis*. C,D, *L. loshii*. E, *N. schlehanii*. F, *N. jongmansii*. G, *N. rectinervis*. After Laveine (1967). All × 3.

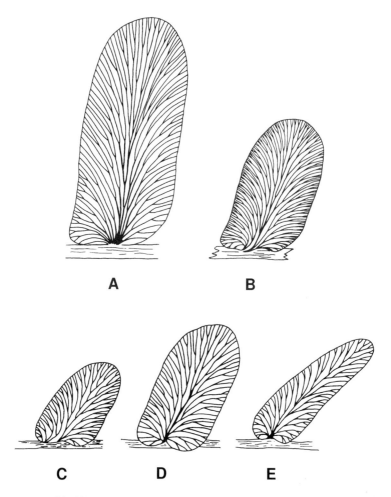

TEXT-FIG. 54. *Neuropteris* pinnules. A, *N. flexuosa*, B, *N. ovata*. C, *N. obliqua*. D, *N. dussartii*. E, *N. hollandica*. A after Crookall (1957), B-E after Laveine (1967). All × 3.

24. Average pinnule length > 20 mm.....................*Neuropteris jongmansii*
 (Pl. 18, fig. 4; Text-fig. 55B)
    Average pinnule length < 20 mm.......................................................25
25. Pinnules subfalcate to rounded, with thick midvein ............................
 .................................................................*Laveineopteris tenuifolia*
 (Pl. 17, fig. 5; Text-fig. 53A)

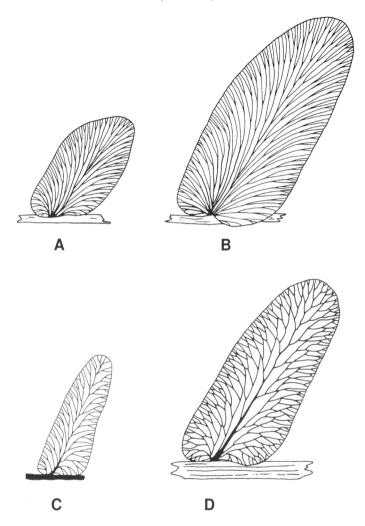

TEXT-FIG. 55. *Neuropteris* and *Reticulopteris* pinnules. A, *N. heterophylla*. B, *N. jongmansii*. C, *N. semireticulata*. D, *R. muensteri*. After Laveine (1967). All × 3.

  Pinnules more slender and triangular, with a finely marked midvein and oblique lateral veins ..................................*Neuropteris hollandica*
(Text-fig. 54E)
26. Pinnules 20–40 mm long, subtriangular, with acuminate apex ........28
  Pinnules rounded to linguaeform, 13–20 mm long............................27

27. Pinnules consistently oval to linguaeform, rarely subtriangular; only fused to rachis high in pinna......................*Neuropteris heterophylla* (Pl. 16; Text-fig. 55A)
    Pinnules more variable in shape, frequently becoming broadly attached to the rachis................................................................................28
28. Lateral veins often slightly flexuous but never pseudoanastomosed.... ..........................................................................*Neuropteris obliqua* (Text-fig. 54C)
    Lateral veins frequently pseudoanastomosed or fully anastomosed..... ............................................................................................................29
29. Lateral veins mainly pseudoanastomosed..*Neuropteris semireticulata* (Pl. 19, fig. 1; Text-fig. 55C)
    Lateral veins exclusively anastomosed..........*Reticulopteris muensteri* (Text-fig. 55D)
30. Large linguaeform pinnules with surface covered by short hairs......... .................................................................*Macroneuropteris scheuchzeri* (Pl. 18, fig. 2; Text-fig. 50A)
    Pinnules with little or no evidence of hairs on the surface, and often with a basiscopic lobe (auricle).........................................................31
31. Large, often subtriangular pinnules, strongly decurrent on acroscopic side, and with the midvein entering the pinnule towards the acroscopic side....................................*Macroneuropteris macrophylla* (Text-fig. 50B)
    Rounded to linguaeform pinnules, rather swollen acroscopically; midvein lying more centrally along pinnule..........................................32
32. Pinnules usually > 20 mm long, and with a large lanceolate apical pinnule..................................................................*Neuropteris flexuosa* (Pl. 22, fig. 1; Text-fig. 54A)
    Pinnules usually < 20 mm long, and with a smaller, usually rather oval apical pinnule .................................................*Neuropteris ovata* (Text-fig. 54B)
33. Lateral veins broadly arched and non-anastomosed........................34
    Lateral veins anastomosed ........................................................36

---

EXPLANATION OF PLATE 18

Figs 1, 2. *Macroneuropteris scheuchzeri*. No. 3 Llantwit Coal (Westphalian D); old tip near Beddau, Mid-Glamorgan, × 1. 1, NMW 92.20G7; incised pinnule-form ('*Odontopteris lindleyana*'). 2, NMW 92.20G6; normal pinnule form.

Fig. 3. *Paripteris pseudogigantea*. NMW 86.101G3; upper Langsettian; Cattybrook Claypit, near Almondsbury, Avon, × 2.

Fig. 4. *Neuropteris jongmansii*. BGS, Kidston Collection, 4521; four feet above the Bassey Mine (Bolsovian); Cobridge Marl Pit, near Cobridge Station, north Staffordshire, × 1.

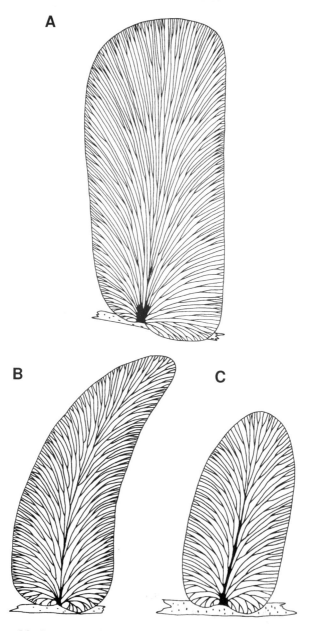

TEXT-FIG. 56. *Paripteris* pinnules. A, *P. linguaefolia*. B, *P. gigantea*. C, *P. pseudogigantea*. After Laveine (1967). All × 3.

## Key to Group E

34. Mainly subfalcate pinnules .................................*Paripteris gigantea*
(Text-fig. 56B)
Linguaeform or subtriangular pinnules ................................................35
35. Weakly developed midvein, extending for < 0·5 pinnule length ..........
.............................................................*Paripteris linguaefolia*
(Text-fig. 56A)
Strongly developed midvein, extending for > 0·5 pinnule length ........
.........................................................*Paripteris pseudogigantea*
(Pl. 18, fig. 3; Text-fig. 56C)
36. Mainly linguaeform pinnules, < 20 mm long; fairly isodiametric vein meshes, except near midvein ...............................*Linopteris bunburii*
(Text-fig. 57C)
Pinnules mainly > 20 mm long ...........................................................37
37. Mainly slender, subfalcate pinnules .............*Linopteris neuropteroides*
(Text-fig. 57A)
Broader, mainly linguaeform pinnules .......*Linopteris subbrongniartii*
(Text-fig. 57B)
38. Midvein absent or very poorly developed ..........................................39
Prominent midvein .............................................................................40
39. Elongate pinnules oblique to rachis, and often deeply constricted on acroscopic side; hairs visible on pinnule surface; nervation dense ......
...........................................................*Macroneuropteris scheuchzeri*
(Pl. 18, fig. 1; Text-fig. 50A)
Mainly squat pinnules; no hairs visible on pinnule surface; wide nervation ........................................................*Odontopteris cantabrica*
(Pl. 19, fig. 3; Text-fig. 58A)
40. Very elongate, thin-limbed pinnules, usually at least 5 times longer than broad. Very prominent midvein but finely marked lateral veins. Nervation density < 15 veins per cm ...............*Desmopteris longifolia*
(Text-fig. 58B)
Shorter pinnules usually < 5 times longer than broad. If more elongate, then pinnules have a relatively thicker limb. Nervation density > 15 veins per cm ...................................................................41
41. Pinnules fairly symmetrical about midvein, and mainly inserted at right-angles to rachis; lateral veins usually oblique to pinnule margin .........42
Pinnules somewhat asymmetrical about midvein, and mainly inserted obliquely to rachis; lateral veins meet pinnule margin at about right-angles ...................................................................................................43
42. Subrectangular to linguaeform pinnules, averaging > 10 mm long
.......................................................*Callipteridium jongmansii*
(Text-fig. 58C)
Subtriangular to linguaeform pinnules, averaging < 10 mm long ........
................................................................*Callipteridium armasii*
(Text-fig. 58D)

Key to Group E     121

43. Nervation not anastomosed.................................................44
    Nervation anastomosed......................................................55
44. Lateral veins unforked or once forked....................................45
    Lateral veins mostly at least once-forked, only very rarely simple..46
45. Very elongate, lanceolate pinnules, often > 20 mm long; broad midvein; forked veins more common than simple veins ..................
    .................................................................*Alethopteris bertrandii*
    (Text-fig. 59A)
    Short, linguaeform pinnules, usually < 20 mm long; narrower midvein; predominantly simple lateral veins........................................
    .................................................................*Alethopteris robusta* var. *longipinnata*
    (Text-fig. 60C)
46. Veins widely forked, sometimes pseudoanastomosing...................47
    Veins narrowly forked........................................................48
47. Very slender pinnules, usually < 3 mm wide...*Alethopteris decurrens*
    (Pl. 21, fig. 3; Text-fig. 61C)
    Subrectangular pinnules, 3–6 mm wide. Widest pinnules have length : breadth ratio 4 to 5..............................................*Alethopteris davreuxii*
    (Text-fig. 61D)
    Biconvex pinnules, 5–8 mm wide. Narrowest pinnules have length : breadth ratio 1·5 to 2·0..........................................*Alethopteris serlii*
    (Pl. 21, fig. 1; Text-fig. 59B)
48. Apical pinnules relatively short and partially fused to the adjacent lateral pinnules; nervation density usually < 40 veins per cm..........49
    Apical pinnules relatively large and constricted at base; nervation density usually > 40 veins per cm................................................53
49. Pinnules biconvex or subtriangular........................................50
    Pinnules more or less parallel sided.......................................52
50. Pinnules subtriangular............................................*Alethopteris valida*
    (Text-fig. 61B)
    Pinnules biconvex..............................................................51

---

EXPLANATION OF PLATE 19

Fig. 1. *Neuropteris semireticulata*. NMW 87.20G78; upper Duckmantian; Howgill Head Quarry, Whitehaven, Cumbria, × 4.

Fig. 2. *Alethopteris grandinioides* var. *grandinioides*. NMW 75.41G4; Pretoria Coal (Westphalian D); Cwm-nant-llwyd, near Pontardawe, West Glamorgan, × 2.

Fig. 3. *Odontopteris cantabrica*. NMW 75.41G101; Coalbrook Coals (Cantabrian); Old Coalbrook Colliery, Grovesend, West Glamorgan, × 2.

Fig. 4. *Lonchopteris rugosa*. NMW; upper Langsettian; Cattybrook Claypit, near Almondsbury, Avon, × 1.

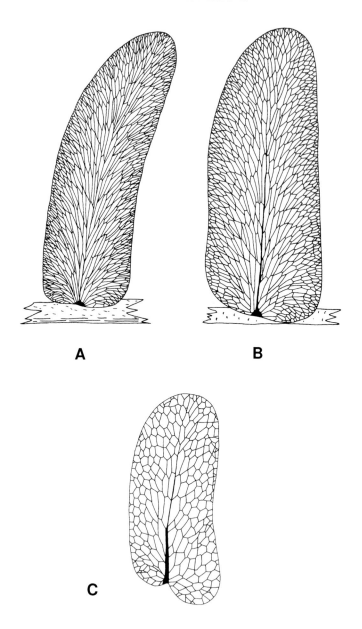

TEXT-FIG. 57. *Linopteris* pinnules. A, *L. neuropteroides*. B, *L. subbrongniartii*. C, *L. bunburii*. A,B after Laveine (1967), C after Crookall (1957). All × 3.

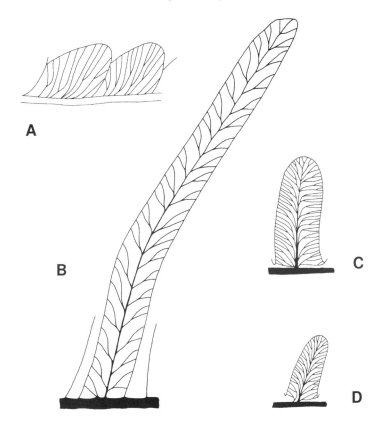

TEXT-FIG. 58. A, *Odontopteris cantabrica*. B, *Desmopteris longifolia*. C, *Callipteridium jongmansii*. D, *C. armasii*. A after Zodrow (1985), B after Buisine (1961), C,D after Laveine *et al.* (1977). All × 3.

51. Veins flexuous, sometimes pseudoanastomosed. Nervation density > 30 veins per cm ...................................................*Alethopteris serlii*
(Pl. 21, fig. 1; Text-fig. 59B)
Veins broadly arched and non-flexuous. Nervation density < 30 veins per cm..........................*Alethopteris grandinioides* var. *grandinioides*
(Pl. 19, fig. 2; Text-fig. 60A)
52. Slender pinnules usually < 5 mm wide, with a bluntly acuminate apex; pinnule limb thick and vaulted; pinnule inserted near to right-angles to rachis; broad midvein ....................*Alethopteris lesquereuxii*
(Pl. 21, fig. 2; Text-fig. 60D)

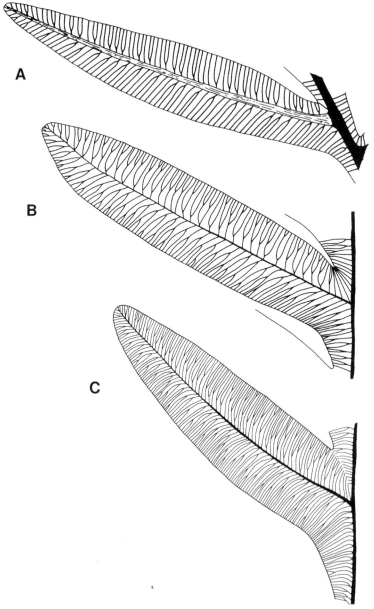

TEXT-FIG. 59. *Alethopteris* pinnules. A, *A. bertrandii*. B, *A. serlii*. C, *A. lancifolia*. A after Buisine (1961), B after Wagner (1968), C after Wagner (1961). All × 3.

Key to Group E 125

Broader pinnules, usually > 5 mm wide, with a broadly rounded apex; pinnule limb rather thinner and less vaulted; pinnule inserted obliquely to rachis; narrower midvein ............................................................................ ....................*Alethopteris grandinioides* var. *subzeilleri* (Text-fig. 60B)
53. Fully developed pinnules subtriangular ............*Alethopteris lancifolia* (Text-fig. 59C)
Fully developed pinnules sublinear to biconvex ..............................54
54. Pinnules high in pinna usually squat and biconvex, and fairly symmetrical about midvein ............................*Alethopteris lonchitica* (Text-figs 60E)
Pinnules high in pinna usually more slender, sublinear or tending to subtriangular; midvein clearly enters acroscopic side of pinnule......... ..............................................................................*Alethopteris urophylla* (Pl. 20; Text-fig. 61A)
55. Small, slender pinnules; very fine nervation, with vein meshes usually < 1 mm long ...............................................*Lonchopteris petitii* (Text-fig. 62B)
More robust pinnules; vein meshes usually > 1 mm long ...............56
56. Pinnules with bluntly acuminate apex, and inserted at about rightangles to rachis, except towards pinna apex, where they become a little oblique; nervation density > 35 veins per cm; vein meshes more or less isodiametric ...............................................*Lonchopteris rugosa* (Pl. 19, fig. 4; Text-fig. 62C)
Pinnules usually with broadly rounded apex, and lying obliquely to rachis; nervation density < 35 veins per cm; vein meshes elongate .... ....................................................................*Lonchopteris eschweileriana* (Text-fig. 62A)
57. Large, irregularly incised pinnules > 50 mm long ......*Aphlebia crispa* (Text-fig. 63)
Smaller, more regularly lobed pinnules < 50 mm long ...................58
58. Pinnules usually > 5 mm long..........................................................59
Pinnules usually < 5 mm long..........................................................90
59. Pinnules with constricted base .......................................................60
Pinnules broadly fused to rachis .....................................................82
60. Pinnules digitate .............................................................................61
Pinnules with broader, mostly rounded, subtriangular or subrectangular lobes...............................................................................63
61. Pinnule segments slender ...............................................................62
Pinnule segments broader, with blunt apices. Pinnules mostly > 10 mm long ....................................................*Palmatopteris sturii* (Text-fig. 64C)
62. Pinnules mostly > 10 mm long .......................*Palmatopteris furcata* (Pl. 24, fig. 2; Text-fig. 64A)

# Key to Group E

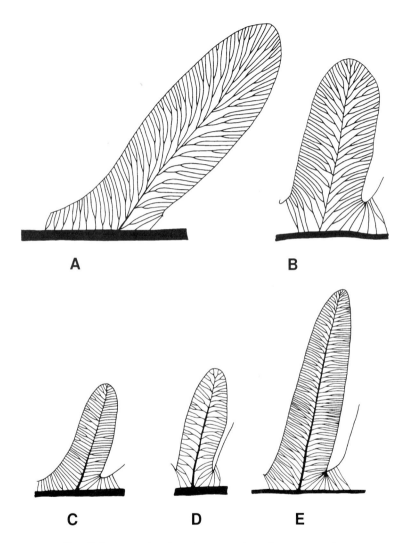

TEXT-FIG. 60. *Alethopteris* pinnules. A, *A. grandinioides* var. *grandinioides*. B, *A. grandinioides* var. *subzeilleri*. C, *A. robusta* var. *longipinnata*. D, *A. lesquereuxii*. E, *A. lonchitica*. After Wagner (1968). All × 3.

EXPLANATION OF PLATE 20

*Alethopteris urophylla*. NMW 87.20G144; Middle Coal Measures (?upper Duckmantian); Howgill Head Quarry, Whitehaven, Cumbria, × 1.

## Key to Group E

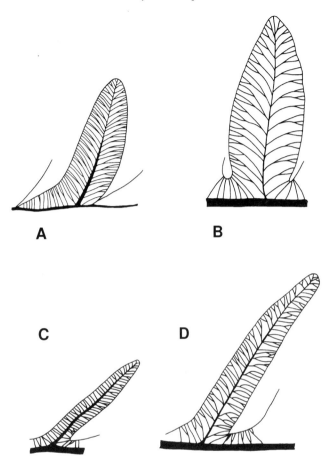

TEXT-FIG. 61. *Alethopteris* pinnules. A, *A. urophylla*. B, *A. valida*. C, *A. decurrens*. D, *A. davreuxii*. A,B after Crookall (1955), C, D after Buisine (1961). All × 3.

---

EXPLANATION OF PLATE 21

Fig. 1. *Alethopteris serlii*. NMW 35.594G42; Farrington Formation (upper Westphalian D); Kilmersdon Colliery Tip, Radstock, Avon, × 1.

Fig. 2. *Alethopteris lesquereuxii*. NMW, David Davies Collection 7938; No. 3 Llantwit Seam (Westphalian D); Bedau, Mid-Glamorgan, × 1.

Fig. 3. *Alethopteris decurrens*. NMW 84.27G250; Coal Measures (Westphalian, probably Langsettian); County Durham, × 1.

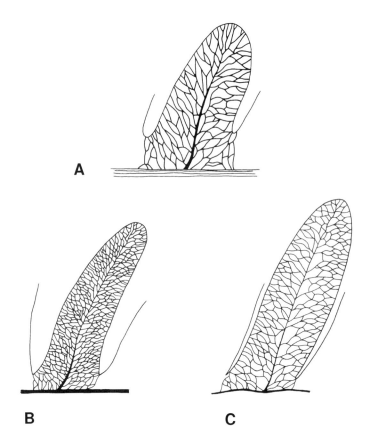

TEXT-FIG. 62. *Lonchopteris* pinnules. A, *L. eschweileriana*. B, *L. petitii*. C, *L. rugosa*. After Buisine (1961). All × 3.

  Pinnules mostly < 10 mm long....................*Palmatopteris geniculata*
             (Pl. 25, fig. 3; Text-fig. 64B)
63. Pinnules at *c*. 45° to rachis..............................*Sphenopteris macilenta*
                (Text-fig. 65B)
  Pinnules at 60–90° to rachis .............................................................64
64. Pinnule lobes with a more or less dentate margin. Midvein present but of equal prominence to lateral veins..........................................65
  Pinnule lobes with smooth margin. No midvein...............................66

TEXT-FIG. 63. *Aphlebia crispa*. After Crookall (1976), × 3.

65. Shallowly lobed pinnules. Rachis only with longitudinal striae............
    ....................................................................*Fortopteris latifolia*
    (Text-fig. 71B)
    Deeply lobed pinnules. Rachis with both longitudinal striae and
    transverse bars.................................................*Sphenopteris andraeana*
    (Text-fig. 65A)
66. Pinnules usually with fairly rounded, symmetrical lobes and rounded
    apex. Drip–tip never present at pinnule/pinna apex. Pinnule surface
    often finely striate. Veins either not visible or, when visible, flush
    with pinnule surface, or in very shallow furrow................................67

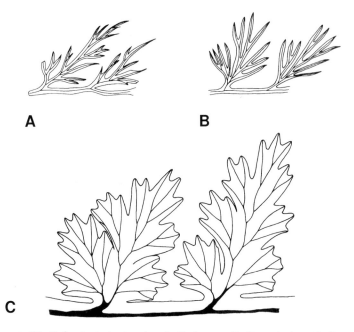

TEXT-FIG. 64. *Palmatopteris* pinnules. A, *P. furcata*. B, *P. geniculata*. C, *P. sturii*. After Kidston (1923–1925). All × 3.

---

EXPLANATION OF PLATE 22

Fig. 1. *Neuropteris flexuosa*. NMW 90.8G5; Farrington Formation (upper Westphalian D); Lower Writhlington Colliery Tip, Radstock, Avon, × 1.

Fig. 2. *Mariopteris nervosa*. NMW 36.429G6; Pennant Formation, Coal Measures (?upper Westphalian); Ynyshir, Mid-Glamorgan, × 1.

Fig. 3. *Polymorphopteris polymorpha*. NMW 90.8G6; Farrington Formation (upper Westphalian D); Lower Writhlington Colliery Tip, Radstock, Avon, × 1.

Pinnules usually with more angular or asymmetrical lobes, and a slender apex which may develop into a drip-tip. Pinnule surface never finely striate. Veins prominently marked and usually in a distinct furrow .................................................................................77
67. Pinnules with smooth surface or showing veins in a shallow furrow ..
.................................................................................................68
Pinnule surface covered by fine striae. If veins are visible, then they are flush with pinnule surface ........................................................73
68. Pinnule lobes cuneiform and elongate .........*Eusphenopteris sauveurii*
(Pl. 26, fig. 3; Text-fig. 66D)
Pinnule lobes rounded to polygonal, and fairly isodiametric ..........69
69. Pinnule lobes strongly rounded and basally constricted. Pinnule lamina flat .......................................................*Eusphenopteris foliolata*
(Text-fig. 68A)
Pinnule lobes angular. Pinnule lamina flat or vaulted ......................70
70. Broad pinnules, often with a flattened limb. Pinnule lobes rounded to subrectangular ....................................*Eusphenopteris neuropteroides*
(Pl. 26, fig. 2; Text-fig. 67)

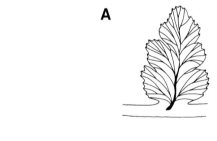

TEXT-FIG. 65. *Sphenopteris* pinnules. A, *S. andraeana*. B, *S. macilenta*. After Boureau and Doubinger (1975). Both × 3.

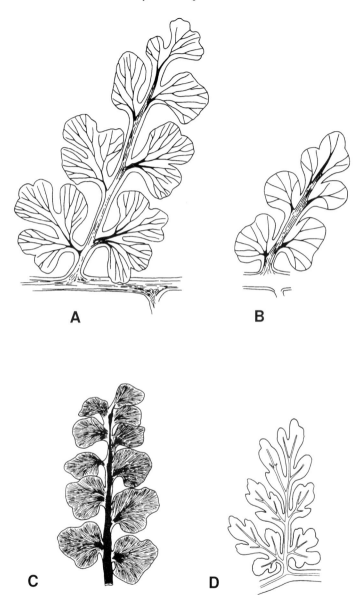

TEXT-FIG. 66. *Eusphenopteris* pinnules. A, *E. obtusiloba*. B, *E. trigonophylla*. C, *E. striata*. D, *E. sauveurii*. After van Amerom (1975). All × 3.

TEXT-FIG. 67. *Eusphenopteris neuropteroides*. After van Amerom (1975), × 3.

    More slender pinnules with a vaulted limb. Pinnule lobes rounded to subtriangular..................................................................................71
71. Pinnules usually > 5 mm long with poorly marked veins......................
..................................................................*Eusphenopteris trifoliolata*
(Pl. 26, fig. 1; Text-fig. 69B)
    Pinnules usually < 5 mm long with strongly marked veins..............72
72. Deeply incised pinnules with distinct petiole. Pinnule limb moderately vaulted.............................................*Eusphenopteris obtusiloba*
(Text-fig. 66A)
    Shallowly incised, subtriangular pinnules with only a short or no petiole. Pinnule limb distinctly vaulted.*Eusphenopteris trigonophylla*
(Text-fig. 66B)
73. Pinnules usually > 5 mm long..........................................................74
    Pinnules usually < 5 mm long..........................................................75
74. Fleshy pinnules with a finely striate surface and distinct petiole. Rachis relatively thin, giving the pinnae a lax appearance .................
.....................................................................*Eusphenopteris grandis*
(Text-fig. 69A)

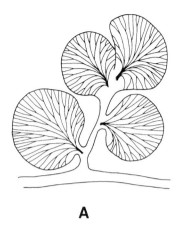 

TEXT-FIG. 68. A, *Eusphenopteris foliolata*. B, *E. scribanii*, × 3. After van Amerom (1975). Both × 3.

   Flat pinnules with a smooth surface, and petioles absent or poorly developed. Relatively thick rachis, giving the pinnae a robust appearance ..................................................... *Eusphenopteris scribanii*
(Text-fig. 68B)
75. Vaulted pinnules, usually < 2 mm long ..... *Eusphenopteris hollandica*
(Text-fig. 69C)
   Vaulted or flat-limbed pinnules, usually > 2 mm long ..................... 76
76. Rounded, vaulted pinnules with smooth surface. Rachis smooth or with a central, longitudinal furrow .......... *Eusphenopteris nummularia*
(Pl. 25, fig. 1; Text-fig. 69D)
   Slightly more subrectangular pinnules with a striate surface. Rachis with longitudinal striae and transverse bars ..... *Eusphenopteris striata*
(Text-fig. 66C)
77. Pinnules which become up to 10 mm long before becoming deeply lobed. Nervation density > 20 veins per cm ...................................... 78
   Pinnules which become deeply lobed when > 5 mm long. Nervation density < 20 veins per cm ................................................................ 79
78. Pinnules with thin limb ........................... *Karinopteris grandepinnata*
(Text-fig. 70E)
   Pinnules with thick limb ...................................... *Karinopteris robusta*
(Text-fig. 71C)
79. Pinnules with thick limb. Veins parallel near pinnule margin............... .......................................................................... *Karinopteris daviesii*
(Text-fig. 70B)

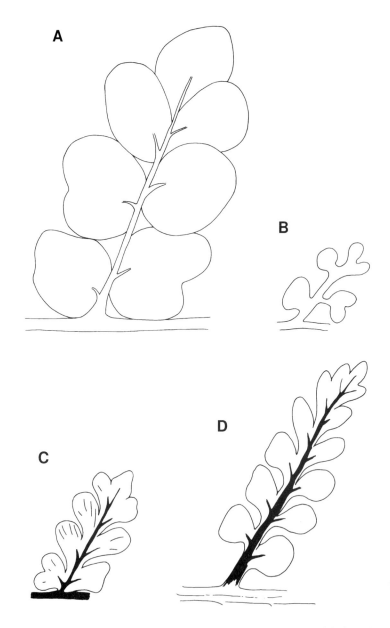

TEXT-FIG. 69. *Eusphenopteris* pinnules. A, *E. grandis*. B, *E. trifoliolata*. C, *E. hollandica*. D, *E. nummularia*. After van Amerom (1975). All × 3.

# Key to Group E

Pinnules with thin limb. Veins slightly fanned near pinnule margin ... ...80
80. Pinnules with rounded to obtuse lobes ..........*Karinopteris soubeiranii*
(Text-fig. 70A)
Pinnules with acute lobes ...81
81. Very strongly marked veins. Prominent hairs usually visible on abaxial surface of pinnules ......*Karinopteris nobilis*
(Text-fig. 70D)
Thinly marked veins. No hairs visible on pinnule surface ......
......*Karinopteris acuta*
(Pl. 24, fig. 1; Text-fig. 70C)
82. Elongate, linguaeform pinnules, < 5 mm wide and with length : breadth ratio 3 or more. Lateral margins of pinnules regularly lobed for much if not all of length. Lobes small and with rounded apices ...
......117
Squatter, usually larger pinnules. Pinnule lobes larger and more angular ...83
83. Pinnules with markedly vaulted lobes. Basiscopic lobe not markedly enlarged relative to acroscopic lobe. Midvein absent or poorly developed......*Dicksonites plueckenetii*
(Text-fig. 72C)
Pinnule limb often vaulted, but individual lobes are not. Basiscopic lobes enlarged relative to acroscopic lobe. Midvein usually well developed, except in smallest pinnules ...84
84. Rachises stout (R1: 4–>9 mm; R2: 3–8 mm; R3: 2–6 mm) ......85
Rachises more slender ...86
85. Pinnules subrectangular with rounded to obtuse apices, and a length : breadth ratio > 2 ......*Mariopteris lobatifolia*
(Text-fig. 71H)
Pinnules narrowly subtriangular with acute to obtuse apices, and a length : breadth ratio 2 to 3·5......*Mariopteris muricata*
(Pl. 23; Text-fig. 71E)
86. Prominent epidermal hairs visible on pinnule surface ...87
No epidermal hairs visible on pinnule surface ...88
87. Robust pinnules, generally squat and subtriangular. Pinnule margin smooth or shallowly lobed. When present, pinnule lobes rounded ......
......*Mariopteris hirta*
(Text-fig. 71F)
Smaller, slenderer pinnules, with a sublanceolate form. Pinnule margin usually with acuminate lobes ......*Mariopteris hirsuta*
(Text-fig. 71A)
88. Pinnules with vaulted limb and strongly marked veins. Low vein density......*Mariopteris nervosa*
(Pl. 22, fig. 2; Text-fig. 71G)

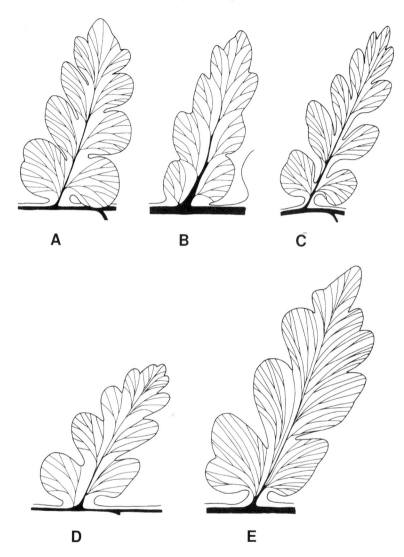

TEXT-FIG. 70. *Karinopteris* pinnules. A, *K. soubeiranii*. B, *K. daviesii*. C, *K. acuta*. D, *K. nobilis*. E, *K. grandepinnata*. After Boersma (1972). All × 3.

---

EXPLANATION OF PLATE 23

*Mariopteris muricata*. NMW 92.20G5; Warren House Coal (Duckmantian); Conney Warren Opencast Pit, near Wakefield, West Yorkshire, × 2.

Key to Group E 143

Pinnules with thinner, flatter limb, and more finely marked veins. Greater vein density ..................................................................89
89. Pinnules with smooth margin, and a rounded to obtuse apex. Length : breadth ratio < 1·5 ................................*Mariopteris sauveurii*
(Text-fig. 71D)
Pinnules with at least one (basiscopic) incision, and an acute to obtuse apex. Length : breadth ratio 1·5 to 2·5 ....*Mariopteris muricata*
(Pl. 23; Text-fig. 71E)
90. Pinnules mostly with thick limb. Broader rachises show longitudinal striae, transverse bars or lozenge-shaped markings ........................91
Pinnules mostly with thinner limb. Rachises show only longitudinal striae ................................................................................93
91. Pinnules deeply incised, except when small (< 1·5 mm long). Rachis with transverse bars or lozenge-shaped markings..............................
......................................................................*Lyginopteris hoeninghausii*
(Text-fig. 72A)
Pinnules with smooth margin or only shallowly incised ................92
92. Very thick-limbed, vaulted pinnules, with lingaueform shape. Pinnule lobes exclusively broadly rounded. Midvein set in deep furrow. Lateral veins closely spaced. Rachis only with longitudinal striae ..........................................................*Margaritopteris conwayii*
(Text-fig. 72D)
Thinner-limbed pinnules with a more triangular aspect. Pinnule lobes more angular. Midvein less prominent and lateral veins widely forked. Rachis with transverse bars or lozenge-shaped marks ............
......................................................................*Lyginopteris baeumleri*
(Text-fig. 72B)
93. Penultimate rachis very broad, almost as wide as the ultimate pinna. Ultimate pinnae parallel-sided for most of length, bearing small, squat pinnules ..................................................................94
Penultimate rachis much narrower than the width of the ultimate pinna. Ultimate pinnae usually tapered for much of length ...........96
94. Subtriangular pinnules with very shallowly lobed margins. Acroscopic pinnule of each pinna elongate and sickle-shaped ..........
......................................................................*Corynepteris similis*
(Text-fig. 73A)

EXPLANATION OF PLATE 24

Fig. 1. *Karinopteris acuta.* NMW 86.101G23; upper Langsettian; Cattybrook Claypit, near Almondsbury, Avon, × 1.
Fig. 2. *Palmatopteris furcata.* NMW 33.187G1; Nine Foot Seam (Duckmantian); Cymmer Colliery, Porth, Mid-Glamorgan, × 1.

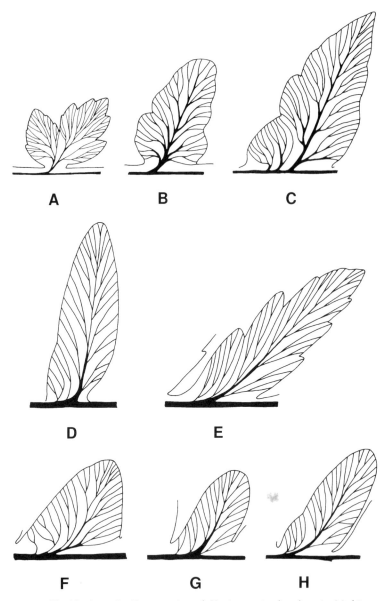

TEXT-FIG. 71. *Mariopteris*, *Fortopteris* and *Karinopteris* pinnules. A, *M. hirsuta*. B, *F. latifolia*. C, *K. robusta*. D, *M. sauveurii*. E, *M. muricata*. F, *M. hirta*. G, *M. nervosa*. H, *M. lobatifolia*. A,B, F, H after Danzé-Corsin (1953), C-E, G after Boersma (1972). All × 3.

# Key to Group E

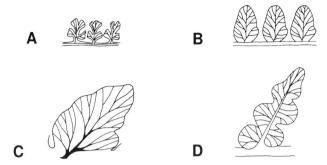

TEXT-FIG. 72. A, *Lyginopteris hoeninghausii*. B, *L. baumleri*. C, *Dicksonites plueckenetii*. D, *Margaritopteris conwayi*. A,B after Kidston (1923–1925), C after Danzé-Corsin (1953), D after Laveine *et al.* (1977). All × 3.

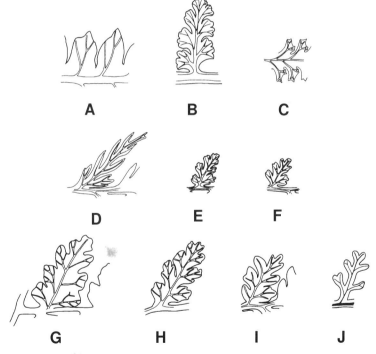

TEXT-FIG. 73. A, *Corynepteris similis*. B, *C. coralloides*. C, *C. angustissima*. D, *Sphenopteris sewardii*. E, *Hymenophyllites quadridactylites*. F, *Sturia amoena*. G, *Sphenopteris rutaefolia*. H, *S. selbyensis*. I, *S. schwerinii*. J, *S. souichii*. After Brousmiche (1983). All × 3.

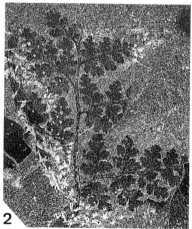

Subrectangular to subcuneiform pinnules with an incised apical margin. Acroscopic pinnule of each pinna digitate ............95
95. Small, more or less parallel-sided pinnules with angular lobes............
............*Corynepteris angustissima*
(Pl. 17, fig. 2; Pl. 27, fig. 1; Text-fig. 73C)
Larger, subcuneiform pinnules with rounded lobes............
............*Corynepteris coralloides*
(Text-fig. 73B)
96. Deeply incised, digitate pinnules with lobes longer than wide ........97
Shallowly lobed pinnules, or with broader, rounded to angular lobes .
............104
97. Small pinnules, usually < 3 mm long............98
Larger pinnules, usually > 3 mm long except near pinna apex ......101
98. Robust rachises, decurrent at base; ultimate rachises > 1 mm wide.....
............*Sphenopteris souichii*
(Text-fig. 73J)
Slender rachises, non-decurrent at base; ultimate rachises < 1 mm wide ............99
99. Pinnules with acute lobes. Sporangia occur over entire pinnule abaxial surface............*Renaultia schatzlarensis*
(Text-fig. 75C)
Pinnules with blunt lobes. Sporangia restricted to pinnule margin ......
............100
100. Pinnae with blunt apex, and with lobed lateral pinnules adjacent to terminal pinnule. No punctae visible on rachides. Sporangia do not extend beyond pinnule margin ........*Hymenophyllites quadridactylites*
(Text-fig. 73E)
Pinnae very gently tapered, and with unlobed lateral pinnules adjacent to terminal pinnule. Rachises bear punctae. Sporangia extend beyond pinnule margin ............*Zeilleria delicatula*
(Text-fig. 74C)
101. Pinnules at < 45° to rachis............102
Pinnules at > 45° to rachis............103
102. Pinnule lobes with rounded apices. Sporangia attached near edge of pinnule but do not extend beyond margin ......*Urnatopteris herbaceae*
(Text-fig. 74A)

EXPLANATION OF PLATE 25

FIG. 1. *Eusphenopteris nummularia.* NMW 83.31G2074; Coal Measures (Westphalian); Rhigos Opencast Colliery, Hirwaun, Mid-Glamorgan, × 1.
Fig. 2. *Renaultia footneri.* NMW 87.20G62; Middle Coal Measures (?upper Duckmantian); Howgill Head Quarry, Whitehaven, Cumbria, × 1.
Fig. 3. *Palmatopteris geniculata.* NMW 86.101G28; Lower Coal Measures (upper Langsettian); Cattybrook Claypit, Almondsbury, Avon, × 1.

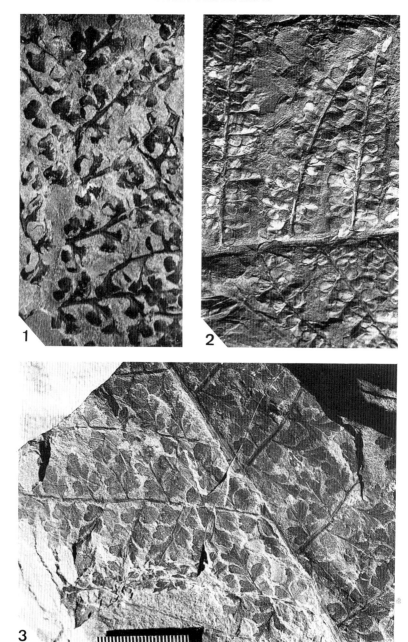

Key to Group E                                              149

Pinnule lobes with acuminate apices. Sporangia unknown ............
......................................................................*Sphenopteris sewardii*
                                                           (Text-fig. 73D)
103. Only pinnules low in pinna are deeply incised. Sporangia large and oval, and attached to specialized pinnules with reduced lamina ..........
..............................................*Crossotheca crepinii*
                                                           (Text-fig. 74G)
    All pinnules exclusively deeply incised. Small sporangia extend beyond pinnule margin ...............................................*Zeilleria frenzlii*
                                                           (Text-fig. 74D)
104. Pinnules with only very shallowly lobed margins and a vaulted limb
    ..................................................................................................105
    Pinnules with deeply incised margins and a flatter limb ..............107
105. Pinnules somewhat triangular and up to 5 mm long. Pinnae parallel-sided with large apical pinnule ........................*Oligocarpia gutbieri*
                                                           (Text-fig. 74E)
    Pinnules round and < 2 mm long. Pinnae tapered with small apical pinnule ......................................................................................106
106. Pinnules deeply incised low in pinna. Clusters of small sporangia attached to unspecialized pinnules ...............*Oligocarpia brongniartii*
                                                           (Text-fig. 74F)
    Pinnules not deeply incised low in pinna. Large sporangia attached to specialized pinnules with reduced limb ...............*Crossotheca crepinii*
                                                           (Text-fig. 74G)
107. Large pinnules, up to 10 mm long ....................................................108
    Small pinnules, never more than 5 mm long ..................................111
108. Pinnules with rounded lobes and often basally confluent low in pinna
    ...................................................................*Renaultia rotundifolia*
                                                           (Text-figs 75A, 76B)
    Pinnules with angular lobes and never basally confluent ..............109
109. Pinnules with subtriangular lobes. Pinnules are lobed very near pinna apex ..........................................................*Renaultia chaerophylloides*
                                                           (Text-fig. 75D)
    Pinnules with parallel-sided lobes. Pinnules not lobed near pinna apex ................................................................................................110

EXPLANATION OF PLATE 26

Fig. 1. *Eusphenopteris trifoliolata*. BMNH V.21085; Barnsley Coal (Duckmantian); Monkton Main Colliery, Barnsley, West Yorkshire, × 2.
Fig. 2. *Eusphenopteris neuropteroides*. BMNH V.21186; Lower Pinchin Coal (Bolsovian); White's Quarry, Ystalyfera, West Glamorgan, × 1.
Fig. 3. *Eusphenopteris sauveurii*. BMNH 52543; Thick Coal (Duckmantian); Tipton, near Dudley, West Midlands, × 1.

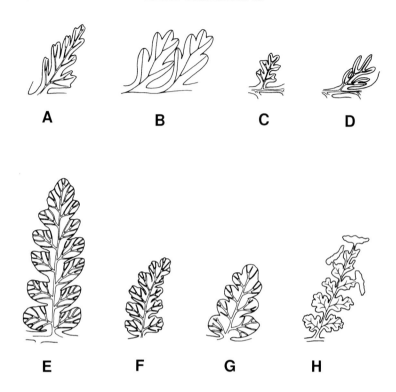

TEXT-FIG. 74. A, *Urnatopteris herbaceae*. B, *Zeilleria hymenophylloides*. C, *Z. delicatula*. D, *Z. frenzlii*. E, *Oligocarpia gutbieri*. F, *O. brongniartii*. G, *Crossotheca crepinii*. H, *Sphyropteris obliqua*. After Brousmiche (1983). All × 3.

110. Pinnules with slender lobes, usually < 1 mm wide. More than 3 veins rarely reach margin of each pinnule lobe ........*Sphenopteris schwerinii* (Text-fig. 73I)
 Pinnules with broad lobes, usually > 1 mm wide. Up to 5 veins may reach margin of each pinnule lobe ..................*Sphenopteris rutaefolia* (Text-fig. 73G)
111. Ultimate pinnae with blunt apex ...........................*Renaultia footneri* (Pl. 25, fig. 2; Text-fig. 75F)
 Ultimate pinnae with gradually tapered apex .....................112
112. Pinnules usually < 2 mm long.......................................113
 Pinnules usually > 2 mm long.......................................114
113. Sporangia distributed over entire pinnule surface ...*Renaultia crepinii* (Text-fig. 75B)

## Key to Group E

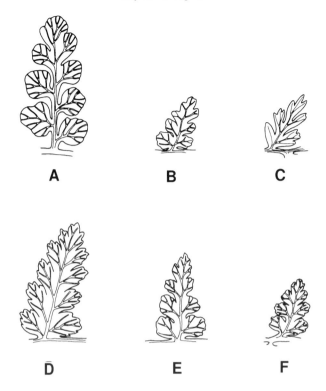

TEXT-FIG. 75. *Renaultia* pinnules. A, *R. rotundifolia*. B, *R. crepinii*. C, *R. schatzlarensis*. D, *R. chaerophylloides*. E, *R. gracilis*. F, *R. footneri*. After Brousmiche (1983). All × 3.

   Sporangia restricted to linear transverse extensions of pinnule apex .. ............................................................................*Sphyropteris obliqua*
                     (Text-fig. 74H)
114. Pinnules become lobed very near pinna apex. Lobes generally longer than broad ........................................................................................115
   Pinnules remain unlobed near pinna apex. Lobes about as long as broad ..........................................................................................116
115. Sporangia distributed over entire pinnule surface ..........*Sturia amoena*
                     (Text-fig. 73F)
   Sporangia restricted to pinnule margin .....*Zeilleria hymenophylloides*
                     (Text-fig. 74B)
116. Pinnules with rounded lobes ....................................*Renaultia gracilis*
                     (Text-fig. 75E)

Pinnules with angular lobes ..........................*Sphenopteris selbyensis*
(Text-fig. 73H)
117. Lobed, pinnatifid pinnules occur abundantly through frond...........118
Lobed, pinnatifid pinnules occur only rarely on frond ...................126
118. Pinnatifid pinnules with elongate, smooth-margined terminal part, more than twice as long as pinnule width. Entire-margined pinnules subtriangular. Sporangia attached singly to abaxial pinnule surface ...
............................................................................................................119
Pinnatifid pinnules with smooth-margined, isodiametric to squat terminal part. Entire-margined pinnules more parallel-sided. Sporangia occur in clusters of (usually) 4, arranged in a row on either side of the midvein, or attached to the ends of the lateral veins and extend beyond pinnule margin........................................................121
119. Pinnatifid pinnules > 5 mm wide, and with more or less pointed apex
.................................................................................*Pecopteris unita*
(Text-fig. 79D)
Pinnatifid pinnules < 5 mm wide, and with a rounded or bluntly acuminate apex ...............................................................................120
120. Lateral veins slightly oblique to pinnule margin. Nervation density *c.* 25 veins per cm .................................................*Pecopteris volkmannii*
(Text-fig. 79B)
Lateral veins meet pinnule margin at about right-angles. Nervation density *c.* 20 veins per cm ...............................*Pecopteris intermedia*
(Text-fig. 77B)
121. Pinnae usually terminated by a well-differentiated, isodiametric, subtriangular apical pinnule.........................................*Lobatopteris vestita*
(Text-fig. 77D)
Pinnae usually terminated by a poorly-differentiated, blunt, rather rounded apical pinnule .................................................................122
122. Pinnatifid pinnules rarely more than 10 mm long before they develop into discrete pinnae .....................................*Lobatopteris micromiltoni*
(Text-fig. 77C)
Pinnatifid pinnules may become > 10 mm long before they develop into discrete pinnae.........................................................................123
123. Pinnatifid pinnules very abundant in frond. Vein density *c.* 35 veins per cm. Penultimate pinnae with blunt apex....................................124

EXPLANATION OF PLATE 27

Fig. 1. *Corynepteris angustissima.* NMW 86.101G30a; upper Langsettian; Cattybrook Claypit, near Almondsbury, Avon, × 1.

Fig. 2. *Cyathocarpus* aff. *arborescens.* BMNH V.894, Kidston Collection; ?Radstock Formation (upper Wesphalian D); Timsbury, Avon, × 1.

TEXT-FIG. 76. A, *Pecopteris plumosa*; NMW 92.20G1; Coal Measures (Westphalian); Warren House Opencast Site, Coney Warren, near Wakefield, West Yorkshire. B, *Renaultia rotundifolia*; BMNH V.1855; Duckmantian(?); Clay Cross, Derbyshire. All × 1.

# Key to Group E 155

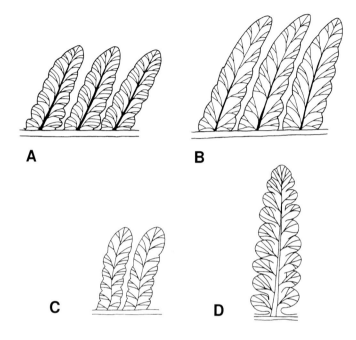

TEXT-FIG. 77. A, *Bertrandia avoldensis*. B, *Pecopteris intermedia*. C, *Lobatopteris micromiltoni*. D, *L. vestita*. A,B after Dalinval (1960). All × 3.

   Pinnatifid pinnules less abundant in frond. Vein density *c.* 25 veins or less per cm. Penultimate pinnae with gradually tapered apex ........125
124. Pinnules covered with fine hairs, often masking the veins. Midvein strongly decurrent at base. Two rows of sori attached to underside of pinnules, one on either side of midvein, halfway between midvein and pinnule margin ...............................................*Lobatopteris miltoni*
                   (Text-fig. 78B)
   No hairs visible on pinnule surface. Midvein rarely decurrent, except in smaller pinnules. Sori attached to ends of lateral veins and extend beyond pinnule margin ........................................*Bertrandia avoldensis*
                   (Text-fig. 77A)
125. Thick midvein extending for most of pinnule length. Widely forking lateral veins, producing vein density < 25 veins per cm......................
   .............................................................................*Pecopteris lobulata*
                   (Text-fig. 78C)

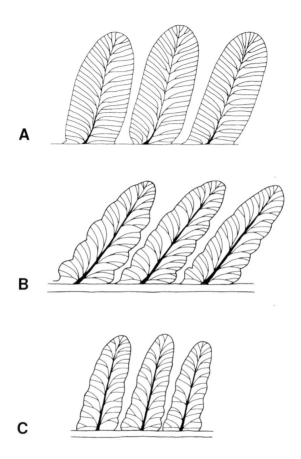

TEXT-FIG. 78. A, *Polymorphopteris polymorpha*. B, *Lobatopteris miltoni*. C, *Pecopteris lobulata*. A after Boureau and Doubinger (1975), B,C after Dalinval (1960). All × 3.

Thinner midvein extending for only 0·75 or less of pinnule length. Narrowly forking lateral veins, producing vein density > 25 veins per cm .................................................................*Pecopteris bourozii* (Text-fig. 79A)

126. Pinnules with biconvex lateral margins, a bluntly acuminate apex and a slightly constricted base. Adjacent pinnules never confluent, except near pinna apex ...........................................................................127

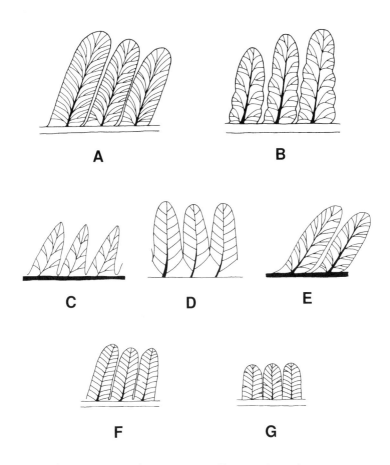

TEXT-FIG. 79. A, *Pecopteris bourozii*. B, *P. volkmannii*. C, *P. plumosa*. D, *P. unita*. E, *P. bucklandii*. F, *Cyathocarpus hemitelioides*. G, *C.* aff. *arborescens*. A-C after Dalinval (1960), D, F,G after Boureau and Doubinger (1975), E after Kidston (1923). All × 3.

    Pinnules with straight lateral margins, rounded or acuminate apex, and broadly attached at base to rachis. Adjacent pinnules may or may not be confluent at base ................................................................. 128
127. Lateral veins meet pinnule margin near to right-angles. Pinna terminals blunt and with a small, well-individualized apical pinnule ............................................................ *Polymorphopteris polymorpha*
(Pl. 22, fig. 3; Text-fig. 78A)

Lateral veins meet pinnule margin obliquely. Pinna terminals gradually tapered ..................................................*Pecopteris bucklandii* (Text-fig. 79E)

128. Pinnules subtriangular ................................................................129
Pinnules nearly parallel-sided ........................................................130

129. Pinnules with pointed apex. Lateral veins usually forked, fairly straight, and meeting pinnule margin at > 45° ......*Pecopteris plumosa* (Text-figs 76A, 79C)
Pinnules with rounded apex. Lateral veins usually simple, straight or curved upwards, meeting pinnule margin at < 45° .....*Pecopteris unita* (Text-fig. 79D)

130. Short pinnules, usually less than twice as long as broad ...............131
Elongate pinnules more than twice as long as broad ....................132

131. Adjacent pinnules joined to each other at base for at least 0·25 of length. Lateral veins straight or slightly curved upwards, reaching pinnule margin at < 45° ..............................................*Pecopteris unita* (Text-fig. 79D)
Adjacent pinnules quite separate, or only joined for < 0·25 of length. Lateral veins straight or curved downwards, reaching pinnule margin at > 45° ................................................*Cyathocarpus* aff. *arborescens* (Pl. 27, fig. 2; Text-fig. 79G)

132. Veins usually once forked, only rarely simple.......................................
.........................................................*Cyathocarpus* aff. *arborescens* (Pl. 27, fig. 2; Text-fig. 79G)
Veins always simple ................................*Cyathocarpus hemitelioides* (Text-fig. 79F)

# KEY TO GROUP F: CIGAR-SHAPED 'CONES' OR STROBILI BORNE SINGLE OR IN A SERIES ALONG A STEM

These are the reproductive organs of lycopsids, calamites, sphenophylls and cordaites. Adpressions of such reproductive organs are notoriously difficult to identify to species level. Those of the lycopsids, belonging to *Flemingites* (cones possessing both megaspores and microspores) and *Lepidostrobus* (cones possessing microspores) cannot be reliably named even to genus without knowledge of their spores (e.g. Chaloner 1953; Thomas 1970*b*, 1987*b*, 1988; Thomas and Dytko 1980; Brack-Hanes and Thomas 1983; Thomas and Blackburn 1987). *Sigillariostrobus*, which represents both microsporangiate and megasporangiate cones of the Sigillariostrobaceae, can be distinguished from the previous two form-genera on gross morphological criteria, but again individual species cannot be recognized without a knowledge of their spores.

Some lycopsid cones fragment to disperse their sporophylls with sporangia attached before the spores are shed. Such sporophylls are included within Group G.

Calamite cones are assigned to form-genera based on the attachment point of the sporangium-bearing stalk, the sporangiophore (e.g. Text-fig. 12B; see Gastaldo 1981 for a fuller discussion). In *Calamostachys*, the sporangiophores are attached to the cone axis half way between the whorls of bracts. In *Palaeostachya*, the sporangiophores are attached to the stalk axis in the axils of the bracts. If the attachment points are indistinguishable, the cones are sometimes referred to *Paracalamostachys*.

Sphenophyll cones are usually found attached to recognizable foliage and are not given names of their own. If isolated, they may be called *Bowmanites* or *Koinostachys*. Sometimes fragments of cones, or groups of isolated sporangia, are found, but these may be keyed out in Group C. Boureau (1964) gives details of the few distinguishable species.

1. Single cones, either borne at the apices of leafy stems or on short stalks that might be attached in whorls to the nodes of *Calamite* stems, or found detached............................................................................2
   A series of small cones (< 20 mm long) borne directly on a leafless axis ..........................*Cordaitanthus* (a genus of both seed-bearing and pollen-producing cones belonging to *Cordaites*)
   (Text-fig. 80)

TEXT-FIG. 80. *Cordaitanthus* sp.; BMNH V.28321; ironstones associated with Crow Coal (lower Duckmantian); Phoenix Brickworks, Crawcrook, Ryton, County Durham, × 1.

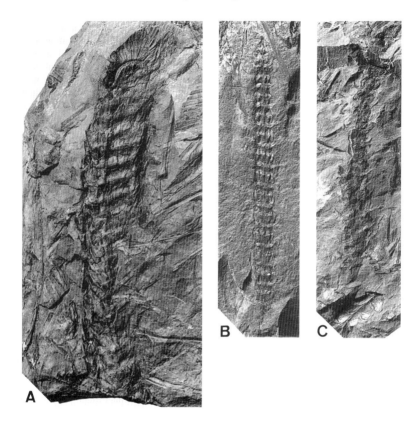

TEXT-FIG. 81. A, *Macrostachya infundibuliformis*; NMW 90.9G4; Farrington Formation (upper Westphalian D); Kilmersdon Colliery Tip, Radstock, Avon. B, *Palaeostachya* sp.; NMW 90.8G8; Farrington Formation (upper Westphalian D); Lower Writhlington Colliery Tip, Radstock, Avon. C, *Calamostachys* sp.; NMW 90.8G9; Farrington Formation (upper Westphalian D); Lower Writhlington Colliery Tip, Radstock, Avon. All × 1.

2. Cone scales borne in spirals. Central axes with spirally arranged attachment points or dehiscence scars of cone scales ......................... 3
Cone scales borne in whorls. Central axis jointed with longitudinal striations as in *Calamites* and *Sphenophyllum* ...................................... 4
3. Cones borne on leafy shoots. Cone scales with smooth margins. (Larger and more leaf-like, spreading scales may be of the *Lepidostrobophyllum* type. These may be identified by cone scales

# Key to Group F

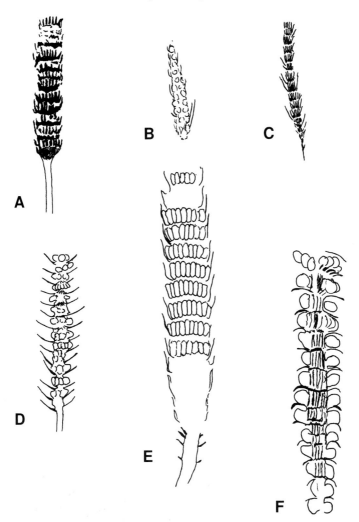

TEXT-FIG. 82. *Palaeostachya* cones. A, *P. paucibracteata*, × 2. B, *P. pedunculata*, × 1. C, *P. minuta*, × 2. D, *P. elongata*, × 1. E, *P. ettingshausenii*, × 1. F, *P. gracillima*, × 1. A after Boureau (1964), B-D after Crookall (1969).

EXPLANATION OF PLATE 28

Fig. 1. *Lepidostrobus* sp. NMW G.1007; Coal Measures (Westphalian); Abercarn, Gwent, × 1.

Fig. 2. *Sigillariostrobus rhombibracteatus*. NMW 92.20G3; Barnsley Thick Coal (Duckmantian); Monkton Main Colliery, Barnsley, South Yorkshire, × 1.

(see Group G) even if the cone has no separate name ..............................
................................................................*Flemingites* or *Lepidostrobus*
(Pl. 28 fig. 1)
Cones borne on leafless stalks. Cone scales with finely-toothed margins
................................................................................*Sigillariostrobus*
(Pl. 28 fig. 2)
4. Cone diameter > 20 mm, axis diameter > 4 mm ......................................
................................................................*Macrostachya infundibuliformis*
(Text-fig. 81A)
Cone diameter < 20 mm, axis diameter < 4 mm ..........................5
5. Two or more sporangia attached to a short stalk (sporangiophore) above each bract...............................................................................6
One to four sporangia positioned above each bract—not on sporangiophores..................................................*Bowmanites*
6. Sporangiophores at right-angles to cone axis midway between bracts
...........................................................................................................7
Sporangiophores oblique to cone axis in axil of bracts.....................12
7. Bracts at least twice as long as internodes (i.e. distance between successive whorls of leaves)..................................................................8
Bracts approximately same length as internodes................................9
8. Bracts approximately twice as long as internode length, spreading out from the cone................................................*Calamostachys germanica*
(Text-fig. 83D)
Bracts approximately 3 times internode length, closely adpressed to cone ........................................................*Calamostachys striata*
(Text-fig. 83A)
9. Cones longer than 200 mm ...................*Calamostachys northumbriana*
(Text-fig. 83E)
Cones shorter than 200 mm ...............................................................10
10. Cones shorter than 60 mm...........................*Calamostachys ramosa*
(Pl. 15, fig. 5; Text-fig. 83B)
Cones longer than 60 mm ..................................................................11
11. Internode length 2–2·5 mm .......................*Calamostachys paniculata*
(Text-fig. 83C)
Internode length 5 mm or more .................*Calamostachys tuberculata*
(Text-fig. 83F)
12. Cones longer than 30 mm ................................................................13
Cones shorter than 30 mm ................................................................14
13. Cones 15 mm long, bracts approximately 1·5 times as long as internode length .......................................................*Palaeostachya minuta*
(Text-fig. 82C)
Cones 30 mm long, bracts just longer than internodes ........................
................................................................*Palaeostachya paucibracteata*
(Text-fig. 82A)

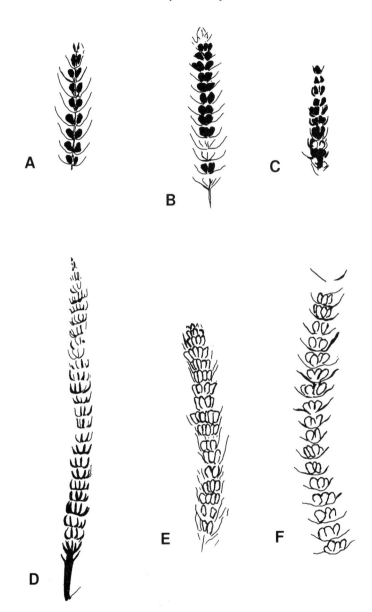

TEXT-FIG. 83. *Calamostachys* cones. A, *C. striata*. B, *C. ramosa*. C, *C. paniculata*. D, *C. germanica*. E, *C. northumbriana*. F, *C. tuberculata*. A-C, E,F after Crookall (1969), D after Weiss (1876). All × 1.

14. Bracts twice as long as internode length ...............................................
     ............................................................*Palaeostachya ettingshausenii*
     (Text-fig. 82E)
     Bracts approximately same length as internodes.............................15
15. Cones 50–110 mm long, sessile or on very short stalks, bracts usually shorter than internode length ..........................*Palaeostachya gracillima*
     (Text-fig. 82F)
     Cones 50–90 mm long, stalk 5–15 mm long, bracts slightly longer than internode length .............................................*Palaeostachya elongata*
     (Text-fig. 82D)
     Cones 23–55 mm long, stalk 4–6 mm long, bracts slightly shorter than internode length ..........................................*Palaeostachya pedunculata*
     (Text-fig. 82B)

# KEY TO GROUP G: DETACHED LYCOPSID CONE-SCALES OR LEAVES

These are cone-scales (sporophylls) from the arborescent lycopsids. They consist of a basal (proximal) pedicel and an apical (distal) leaf-like lamina. The pedicel usually has a large sporangium attached to its upper surface and remnants of the cone axis may sometimes be seen at its proximal end. The size and shape of the distal lamina are the characters on which the species are based. These characters are, unfortunately, very unreliable because of the large amount of variation that can be found within a cone (Boulter 1968). It is also impossible to identify sporophylls that are laterally compressed.

These sporophylls are shed naturally from cones as part of their dispersal mechanism, but sometimes entire cones or fragments of them may be preserved (Thomas 1981). Such specimens will key out in Groups C and F.

1. Long, narrow grass-like leaf, up to 400 mm long and 5 mm wide, with up to 3 longitudinal grooves and/or ridges............................................2
   Triangular, wedge or arrow-shaped cone-scale with distal lamina extending from a shorter, proximal stalk which may bear an ovoidal sporangium............................................................................................3
2. Leaf margins ciliated..................................................*Cyperites ciliatus*
   (Text-fig. 26A)
   Leaf margins smooth............................................*Cyperites bicarinatus*
   (Text-fig. 26B)
3. More or less triangular-shaped sporophyll with incurved sides.............4
   Arrow- or wedge-shaped sporophyll.....................................................5
4. Sporophyll sides markedly incurved and tapering to an acuminate apex, margins ciliate................................*Sigillariostrobus rhombibracteatus*
   (Pl. 28, fig. 2)
   Sporophyll with an out-turned lobe on either side of lamina base, margins smooth..................................*Lepidostrobophyllum hastatum*
   (Text-fig. 84A,C)
5. Pedicel with lateral extensions (alations)—not to be confused with a sporangium..........................................................................................6
   No lateral extensions to the pedicel.....................................................8
6. Sporophyll more or less triangular or arrow-shaped, about 15–25 mm long..........................................................*Lepidostrobophyllum alatum*
   (Text-figs 84D-F,J,K, 85)

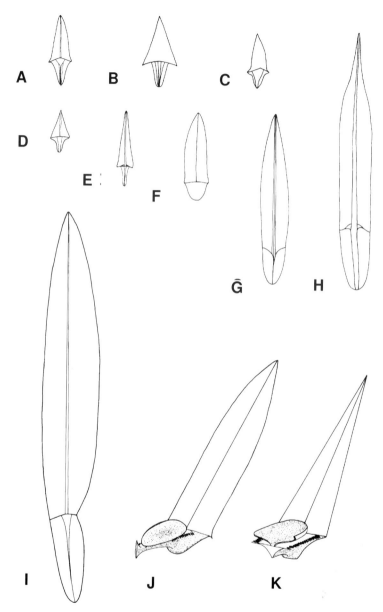

TEXT-FIG. 84. *Lepidostrobophyllum* sporophylls. A, C, *L. hastatum*. B, *L. triangulare*. D-F, J, K, *L. alatum*. G, *L. lanceolatum*. H, *L. acuminatum*. I, *L. majus*. A-C, G-I after Crookall (1966), D-F, J, K after Boulter (1968). All × 1.

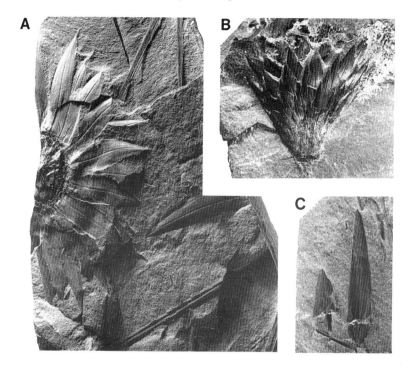

TEXT-FIG. 85. *Lepidostrobophyllum alatum* sporophylls; Farrington Formation (upper Westphalian D); Lower Writhlington Colliery Tip, Radstock, Avon. A, NMW 90.8G. B, BMNH V.60430. C, BMNH V.60432. All × 1.

    Sporophyll wedge-shaped, > 30 mm long ............................................... 7
7. Sporophyll up to *c.* 50 mm long ................*Lepidostrobophyllum alatum*
    (Text-figs 84D-F,J,K, 85)
    Sporophyll *c.* 120 mm long ........................*Lepidostrobophyllum majus*
    (Text-fig. 84I)
8. Sporophyll *c.* 15–25 mm long, triangular with acute apex .....................
    ...........................................................*Lepidostrobophyllum triangulare*
    (Text-fig. 84B)
    Sporophyll longer than 25 mm, wedge-shaped ...................................... 9
9. Sporophyll about 70–90 mm long, gradually broadening to its greatest width near its upper end, then rapidly narrowing to an acuminate apex
    ...........................................................*Lepidostrobophyllum acuminatum*
    (Text-fig. 84H)

Sporophyll about 50–125 mm long, gradually broadening to its greatest width near its middle then gradually tapering to an acute apex ............
...................................................*Lepidostrobophyllum lanceolatum*
(Text-fig. 84G)

# KEY TO GROUP H: SEEDS AND BELL-SHAPED CLUSTERS OF POLLEN SACS

This group includes pteridosperm seeds and pollen-bearing organs, together with isolated cordaite seeds. The latter were borne in life in strobili, which are covered in Group F of this key (as are the cordaite pollen-bearing organs). However, the cordaite seeds are often found as detached organs, which will key-out in this group.

There has been a considerable amount of work done on such organs preserved as coal–ball petrifactions, but the fine detail of their internal structure that such material shows is difficult or impossible to determine from adpressions and casts. Halle (1933) was able to demonstrate some of the detailed structures in pollen-bearing organs preserved as compressions, but it depended on involved laboratory techniques and there have been few subsequent studies of this type in the Coal Measures. Consequently, such seeds and pollen-bearing organs are amongst the least understood of the plant fossils preserved in the Coal Measures adpression assemblages.

The fructifications are almost invariably found detached from their parent plant. The very rare examples of a fructification attached to a frond or a stem are of exceptional scientific importance, providing an important insight into the affinities of the Coal Measures plants.

The best reviews of such fossils published in recent years are by Stockmans and Willière (1961) and Crookall (1976), and the following part of the key is based largely on the latter's work. Important characters for identifying the seeds are their general size, shape and radial symmetry, and the presence or absence of longitudinal ribs or lateral wings. If lateral wings are present, their form in the distal (micropylar) part of the seed is especially important. Identifying the pollen-bearing organs, which were evidently delicate structures in life and thus usually badly distorted during fossilization, can be very difficult. If well enough preserved, they may be identified by their overall shape and size, and the number and distribution of sporangia within the body.

1. Seeds, usually retaining a three-dimensional structure even when compressed..................................................................................2
   Clusters of pollen sacs, which are usually clearly flattened...............34
2. Seeds with two lateral wings ............................................................3
   Seeds without lateral wings .............................................................15

3. Lateral wings with prominent apical 'horns' extending beyond micropyle of seed ............................................................................ 4
   Lateral wings without apical 'horns' ...................................................... 7
4. Seeds longer than broad ........................................................................ 5
   Seeds about as long as broad .................................................................. 6
5. Seed < 20 mm long ............................................................ *Samaropsis bisecta*
   (Text-fig. 86D)
   Seed > 20 mm long ........................................... *Polypterocarpus johnsonii*
   (Text-fig. 86F)
6. Apical horns of lateral wings closely spaced, forming a notch above micropyle ........................................................................ *Cornucarpus acutus*
   (Text-fig. 87B)
   Apical horns of lateral wings widely spaced, forming a wide U-shaped 'valley' above micropyle ........................................... *Cornucarpus arberi*
   (Text-fig. 87A)
7. Longitudinal rib or ribs on surface of seed between wings .................. 8
   Surface of seed smooth between wings ............................................ 11
8. A single rib between lateral wings ...................................................... 9
   More than one rib between lateral wings .......................................... 10
9. Seeds < 10 mm long. Wings join over apex of seed, apparently enclosing micropyle ............................... *Cordaitanthus pseudofluitans*
   (Text-fig. 87F)

EXPLANATION OF PLATE 29

Fig. 1. *Whittleseya* sp. NMW 87.20G141a; Middle Coal Measures (?upper Duckmantian); Howgill Head Quarry, Whitehaven, Cumbria, × 4.

Fig. 2. *Boulaya fertilis*. BMNH V.3355; Ten Foot Ironstone (Duckmantian); Coseley, Dudley, West Midlands, × 1.

Figs 3, 6. *Trigonocarpus parkinsonii*. NMW; Peel Hall Rock (lower Duckmantian); Peel Hall Quarry, near Bolton, Greater Manchester, × 1.

Fig. 4. *T. parkinsonii*. BMNH V.40584; Coal Measures (Westphalian); Newcastle-upon-Tyne, × 1.

Fig. 5. *Samaropsis gutbieri*. BMNH V.1180; Coal Measures (Westphalian); Dudley, West Midlands, × 1.

Fig. 7. *Cordaicarpus cordai*. NMW 86.101G68; Lower Coal Measures (upper Langsettian); Cattybrook Claypit, Almondsbury, Avon, × 2.

Fig. 8. *Aulacotheca hallei*. BMNH V.33388; Ten Foot Ironstones (Duckmantian); Coseley, Dudley, West Midlands, × 1.

Fig. 9. *Holcospermum sulcatum*. BMNH V.241; Coal Measures (Westphalian); Longport, north Staffordshire, × 1.

Fig. 10. *Carpolithus wildii*. BMNH V.21096; Barnsley Coal (Duckmantian); Monkton Main Colliery, Barnsley, South Yorkshire, × 1.

Fig. 11. *Potoniea carpentieri*. BMNH V.33301; Thick Coal (Duckmantian); Dudley, West Midlands, × 1.

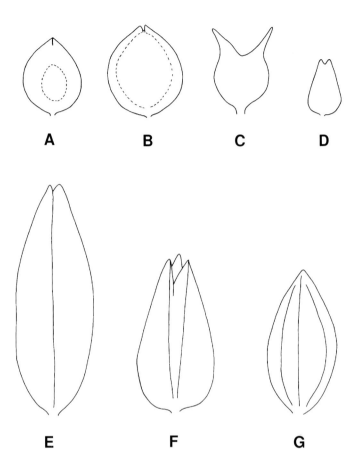

TEXT-FIG. 86. A, *Samaropsis emarginata*. B, *S. gutbieri*. C, *S. crassa*. D, *S. bisecta*. E, *Polypterocarpus anglicus*. F, *P. johnsonii*. G, *P. ornatus*. After Crookall (1976). All × 3.

Seeds > 10 mm long. Wings do not join over apex of seed, and form a notch extending down to micropyle ................... *Cordaitanthus lindleyi* (Text-fig. 87G)
10. Three ribs. Seed > 3 mm long ...................... *Polypterocarpus anglicus* (Text-fig. 86E)
Twelve ribs. Seed < 3 mm long ...................... *Polypterocarpus ornatus* (Text-fig. 86G)

# Key to Group H

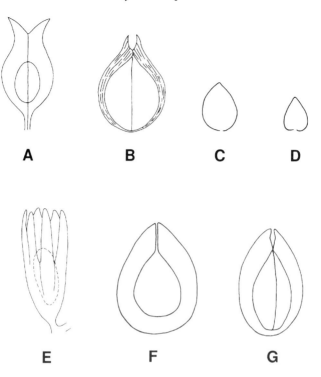

TEXT-FIG. 87. A, *Cornucarpus arberi*. B, *C. acutus*. C, *Cordaicarpus cordai*. D, *C. congruens*. E, *Lagenospermum* sp. F, *Cordaitanthus pseudofluitans*. G, *C. lindleyi*. After Crookall (1976). All × 3.

11. Lateral wing < 0·125 of width of seed.................................................12
    Lateral wing > 0·125 of width of seed.................................................13
12. Seeds usually > 10 mm long with blunt apex, and rounded or flattened base......................................................*Cordaicarpus cordai*
    (Pl. 29, fig. 7; Text-fig. 87C)
    Seeds usually < 10 mm long with pointed apex, and emarginate base ........................................................*Cordaicarpus congruens*
    (Text-fig. 87D)
13. Seeds < 10 mm long with rounded base...................*Samaropsis crassa*
    (Text-fig. 86C)
    Seeds > 10 mm long with emarginate base .........................................14
14. Lateral wings > 0·33 as wide as seed body ......*Samaropsis emarginata*
    (Text-fig. 86A)

176    *British Coal Measures*

    Lateral wings < 0·33 as wide as seed body............*Samaropsis gutbieri*
                                                                   (Pl. 29, fig. 5; Text-fig. 86B)
15. Seeds with elongate apical lobes ........................................................16
    Seeds without apical lobes ...................................................................17
16. Apical lobes > twice length of seed body.................*Gnetopsis anglica*
                                                                                   (Text-fig. 88A)
    Apical lobes < twice length of seed body ..............*Lagenospermum* sp.
                                                                                   (Text-fig. 87E)
17. Seeds with distinct longitudinal ridges or ribs ....................................18
    Seeds without distinct longitudinal ridges, but may have longitudinal striae or irregular suface undulations..................................................26
18. Seeds with > 6 ribs..............................................................................19
    Seeds with 6 or 3 ribs ........................................................................22

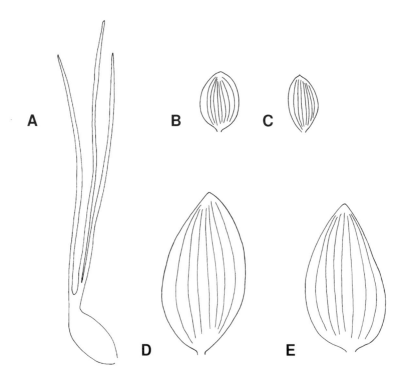

TEXT-FIG. 88. A, *Gnetopsis anglica*. B, *Holcospermum mentzelianum*. C, *H. mammillatum*. D, *H. multistriatum*. E, *H. sulcatum*. After Crookall (1976). All × 3.

19. Seeds usually < 20 mm long .................................................. 20
    Seeds usually > 20 mm long .................................................. 21
20. Seeds usually > 15 mm long, with length : breadth ratio > 2. Ribs more or less straight ............................... *Holcospermum mentzelianum*
    (Text-fig. 88B)
    Seeds usually < 15 mm long, with length : breadth ratio < 2. Ribs slightly flexuous ........................................ *Holcospermum mammillatum*
    (Text-fig. 88C)

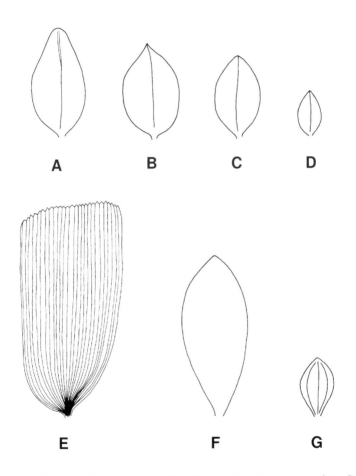

TEXT-FIG. 89. A, *Trigonocarpus oliveriformis*. B, *T. noeggerathii*. C, *T. parkinsonii*. D, *T. candollianus*. E, *Whittleseya* sp. F, *Rhabdocarpus renaultii*. G, *Hexagonocarpus hookeri*. After Crookall (1976). All × 3.

21. Seeds > 35 mm long with obtuse apex. 8 longitudinal ribs. No clear striae between ribs ................................*Holcospermum multistriatum* (Text-fig. 88D)
Seeds < 35 mm long with pointed apex. 9–10 longitudinal ribs. Fine striae between ribs.........................................*Holcospermum sulcatum* (Pl. 29, fig. 9; Text-fig. 88E)
22. Seeds with 6 equally prominent ridges..........*Hexagonocarpus hookeri* (Text-fig. 89G)
Seeds with 3 equally prominent ridges, or with 3 prominent and 3 subsidiary ridges ....................................................................23

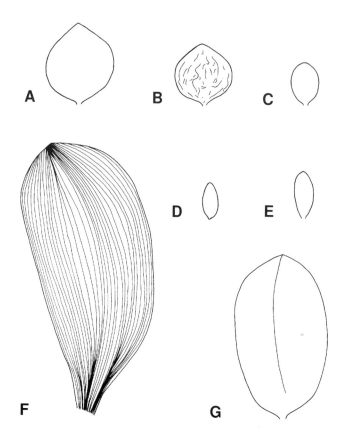

TEXT-FIG. 90. A, *Carpolithus granularis*. B, *C. areolatus*. C, *C. membranaceous*. D, *C. pseudosulcatus*. E, *C. perpusillus*. F, *C. wildii*. G, *C. inflatus*. After Crookall (1976). All × 3.

Key to Group H 179

23. Seeds > 15 mm long ............................................................24
    Seeds < 15 mm long ............................................................25
24. Seeds > 25 mm long ....................................*Trigonocarpus noeggerathii*
    (Text-fig. 89B)
    Seeds < 25 mm long ......................................*Trigonocarpus parkinsonii*
    (Pl. 29, figs 3–4, 6; Text-fig. 89C)
25. Seeds with only 3 prominent ridges and with rounded base. Seed length > 7 mm ........................................*Trigonocarpus candollianus*
    (Text-fig. 89D)
    Seeds with 3 very prominent ridges and another 3 less prominent ridges. Seed length < 7 mm .....................*Trigonocarpus oliveriformis*
    (Text-fig. 89A)
26. Seeds > 20 mm long ..........................................................27
    Seeds < 20 mm long ..........................................................29
27. Seeds with prolonged apex forming micropylar beak .......................
    ...................................................................*Rhabdocarpus renaultii*
    (Text-fig. 89F)
    Seeds with rounded or obtusely pointed apex ...........................28
28. Seeds > 50 mm long with obtusely pointed apex...*Carpolithus inflatus*
    (Text-fig. 90G)
    Seeds < 50 mm long with rounded apex ..................*Carpolithus wildii*
    (Pl. 29, fig. 10; Text-fig. 90F)
29. Seed surface with irregular folds or meshwork ...........................30
    Seed surface smooth ..........................................................32
30  Oval seeds, at least twice as long as broad, with 2 or 3 fold-like undulations .............................................*Carpolithus pseudosulcatus*
    (Text-fig. 90D)
    Oval to subcircular seeds, less than twice as long as broad ..............31
31. Seeds with irregular reticulation on surface ........*Carpolithus areolatus*
    (Text-fig. 90B)
    Seeds with shallow, irregular folds ..........*Carpolithus membranaceous*
    (Text-fig. 90C)
32. Seeds orbicular in outline .................................*Carpolithus granularis*
    (Text-fig. 90A)
    Seeds oval/elliptical in outline .........................................................33
33. Seeds < 3 mm long .............................................*Carpolithus perpusillus*
    (Text-fig. 90E)
    Seeds > 3 mm long ...........................................*Carpolithus membranaceous*
    (Text-fig. 90C)
34. Solid, conical cluster of sporangia, compressed vertically to form a more or less circular disc of carbon. Disc shows wither ridges radiating from a slightly eccentric pedicel, or numerous clusters of 4 sporangia ..............................................................*Potoniea carpentieri*
    (Pl. 29, fig. 11; Text-fig. 91A)

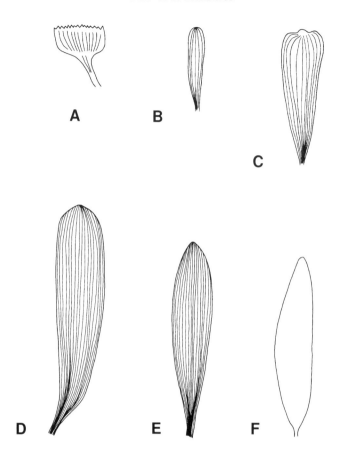

TEXT-FIG. 91. A, *Potoniea carpentieri*. B, *Boulaya praelonga*. C, *B. fertilis*. D, *Aulacotheca hemingwayi*. E, *A. elongata*. F, *A. hallei*. After Crookall (1976). All × 3.

    Hollow cluster of elongate sporangia, forming 'bell'- or 'sausage'-shaped compression, with longitudinal ridges representing the individual sporangia ............................................................................................. 35

35. Cuneate-spatulate compression of a sporangial cluster, 30–60 mm long, with dentate distal margin ................................... *Whittleseya* sp.
                                                                     (Pl. 29, fig. 1; Text-fig. 89E)
    Cylindrical, clavate, elliptical or pyriform sporangial cluster with rounded to bluntly acuminate distal margin ..................................... 36

36. Pyriform to clavate sporangial cluster with irregularly dentate distal margin. Longitudinally striate but with few, if any, longitudinal ribs ....................................................................................................37
    Cylindrical to clavate sporangial cluster with broadly rounded distal margin ....................................................................................38
37. Elongate sporangial cluster, 21–25 mm long and 4–5 mm wide. Not significantly swollen distally...................................*Boulaya praelonga*
    (Text-fig. 91B)
    Squatter sporangial cluster, 14–24 mm long and 8–9 mm wide. Distinctly swollen distally............................................*Boulaya fertilis*
    (Pl. 29, fig. 2; Text-fig. 91C)
38. Sporangial cluster which is somewhat swollen distally, with 9 very prominent ribs around entire circumference ..*Aulacotheca hemingwayi*
    (Text-fig. 91D)
    Cylindrical to elliptical sporangial cluster with 6 or 9 less prominent ribs around entire circumference .......................................................39
39. Six longitudinal ribs..................................................*Aulacotheca hallei*
    (Pl. 29, fig. 8; Text-fig. 91F)
    Nine longitudinal ribs .........................................*Aulacotheca elongata*
    (Text-fig. 91E)

# BIOSTRATIGRAPHY

Plant adpressions have been extensively used for biostratigraphy in the European Coal Measures, and have rivalled the animal fossils in providing detailed correlations; in no other part of the geological column have they played such an important stratigraphical role. Most widely used are fragments of pteridosperm, fern and sphenopsid foliage, from plants growing mainly on the levees. They are particularly useful stratigraphically because there was a steady change in the vegetation with time (see range charts in Text-figures 92–94). This may have been partially triggered by the climatic changes described by Phillips *et al.* (1985). However, the swamp vegetation, which would also have been vulnerable to climatic influence, does not seem to show such marked compositional changes (Phillips 1980); only four or possibly five zones can be recognized in the Westphalian and Stephanian coal ball fossils, while fifteen zone/ subzones can be recognized in the adpression record (Cleal 1991*a*). Additional factors must therefore have been responsible for the more rapid changes in the levee vegetation. Perhaps the most important was the instability of the levees, which were constantly moving with the migration of the river channels, and were regularly subject to slumping and breaching. Such an unstable environment would be ideally suited to invasion by opportunistic species, resulting in a rapidly changing vegetation.

There is little direct evidence of phyletic gradualism in the Coal Measures plant fossil record, one of the few exceptions being the gradual change from *Neuropteris obliqua* to *Reticulopteris muensteri*, documented by Josten (1962; see also Zodrow and Cleal 1993). It is thus tempting to invoke the punctuated equilibrium model of Eldredge and Gould (1972) to explain the temporal dynamics of the vegetation. However, only a small number of palaeoequatorial habitats are represented in the macrofossil record, and we are thus probably seeing a very incomplete picture of the evolutionary history of these plants. For instance, it is known that during at least part of the Westphalian there were extensive conifer-forests surrounding the coal basins, in areas known as extra-basinal habitats (Scott and Chaloner 1983; Lyons and Darrah 1989). The extra-basinal habitats also appear to have supported ginkgoaleans, cycads, peltasperms and even possibly glossopterids, but which rarely entered the areas of sedimentary basins and thus the fossil record, until the Permian (DiMichele and Aronson 1992). There is even evidence that vegetation similar to that preserved in the Lower Carboniferous fossil record, including *Archaeocalamites* and *Sphenopteridium*, was growing in the extra-basinal habitats during the Westphalian, only to return to the basinal habitats in the very late Carboniferous and Permian (Mamay and Bateman 1991;

Mamay 1992). There was clearly a range of vegetational habitats that rarely, if ever, is represented in the macrofossil record, and where many of the phylogenic events that involved the Coal Measures species took place. Another complication is the long-range geographical migration of some of the plants during the Carboniferous, resulting in apparent anomalies in their stratigraphical ranges. A number of species have, for instance, been reported from significantly lower stratigraphical levels in the Ukraine (Fissunenko and Laveine 1984) and China (Laveine *et al.* 1987) compared with western Europe. These patterns can only be recognized by comparing the stratigraphical ranges of different plant and animal fossils with each other in different parts of the world. Unfortunately, such information is not available in many parts of the world; even in much of North America, the ranges of Carboniferous plant fossils have not been properly documented.

Despite all these problems, there is a remarkable consistency in the stratigraphical ranges of many Westphalian plant fossils over large areas of Europe and eastern North America (e.g. Laveine 1977; Cleal 1984*b*; Zodrow and Cleal 1985). We still do not have a clear explanation for this consistency, except for those cases where the phyletic change seems to have taken place within the basinal habitats, such as the *Neuropteris-Reticulopteris* transition noted above. It may be that rapid and geographically widespread environmental changes were taking place on the levees, allowing the sudden influx of pre-adapted plants from adjacent areas. It may reflect genetic drift in a plant lineage, which at a certain point achieved a combination of characters that were particularly suited to the levee habitats and could thus oust the existing vegetation. An extinction of an existing species on the levee, perhaps due to predation or pathological causes, would have provided the opportunity for the immigration of new species. It may even reflect the geographical spread of a species eventually reaching Europe. Whatever the reason, the value of plant fossils for detailed biostratigraphical work in the Upper Carboniferous is now well established (Cleal 1991*a*). In no other part of the geological column have plant macrofossils proved so stratigraphically useful, with subzones representing time intervals of about one million years (compare for instance the position in the Mesozoic—Vakhrameev 1991).

*Historical background*

Much of the stratigraphical work on Upper Carboniferous plant fossils has been done in continental Europe and has been reviewed by Wagner (1984) and Cleal (1991*a*). In Britain, there was some early progress in the field, culminating in the work of Dix (1934), who developed a series of zones modelled on the scheme of Bertrand (1914). Subsequently, however, greater emphasis was placed on the non-marine bivalves and marine bands for correlating the British Coal Measures. The generally held view has been '... that whereas plants give good indication of broad

| Zones | Pecopteris aspera | Lysinopteris hoeninghausii | | Lonchopteris rugosa | | Paripteris linguaefolia | |
|---|---|---|---|---|---|---|---|
| Subzones | Neuralethopteris larischii | Neuralethopteris jongmansii | Laveineopteris loshii | Neuropteris hollandica | Sphenophyllum majus | Neuropteris semireticulata | Laveineopteris rarinervis |
| Neuralethopteris schlehanii | | | | | | | |
| Karinopteris acuta | | | | | | | |
| Lyginopteris baeumleri | | | | | | | |
| Eusphenopteris hollandica | | | | | | | |
| Annularia radiata | | | | | | | |
| Asterophyllites grandis | | | | | | | |
| Sphenophyllum cuneifolium | | | | | | | |
| Paripteris gigantea | | | | | | | |
| Renaultia crepinii | | | | | | | |
| Alethopteris lonchitica | | | | | | | |
| Pecopteris plumosa | | | | | | | |
| Neuropteris obliqua | | | | | | | |
| Alethopteris valida | | | | | | | |
| Lonchopteris eschweileriana | | | | | | | |
| Alethopteris decurrens | | | | | | | |
| Bertrandia avoldensis | | | | | | | |
| Eusphenopteris grandis | | | | | | | |
| Neuralethopteris jongmansii | | | | | | | |
| Lyginopteris hoeninghausii | | | | | | | |
| Asterophyllites charaeformis | | | | | | | |
| Renaultia gracilis | | | | | | | |
| Corynepteris angustissima | | | | | | | |
| Eusphenopteris nummularia | | | | | | | |
| Eusphenopteris trigonophylla | | | | | | | |
| Eusphenopteris obtusiloba | | | | | | | |
| Zeilleria delicatula | | | | | | | |

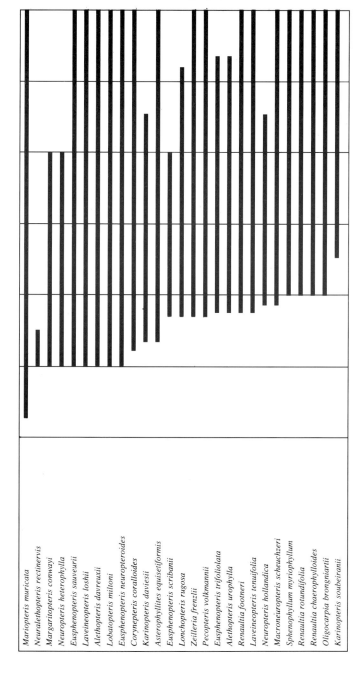

TEXT-FIG. 92. Main stratigraphical ranges of some of the key plant fossil species in the Langsettian to middle Bolsovian.

divisions and of general overseas correlations the ranges of individual species are usually too long to provide the finer divisions given by the bivalves' (Ramsbottom et al. 1978, p. 5). While there is a modicum of truth in this, it tends to debase the valuable contribution that plant macrofossils can make in detailed stratigraphical work, such as the recent studies on the upper Westphalian and lower Stephanian of southern Britain (Wagner and Spinner 1972; Cleal 1978, 1984a, 1984b, 1986b, 1987, 1991b; Zodrow and Cleal 1985).

The most significant development in recent years was the establishment of a new biostratigraphical classification for these fossils by Wagner (1984). Wagner's approach was new, in that his scheme was intended for use throughout much of the Carboniferous palaeoequatorial belt; those of earlier authors tended to be directed towards particular areas (e.g. South Wales by Dix, 1934, Nord-Pas-de-Calais by Bertrand, 1914). Inevitably, Wagner's zones tend to be somewhat longer ranging than those in the earlier schemes, but their wider applicability makes them a sounder basis for establishing inter-coalfield correlations. Wagner's scheme is the basis of the following discussion.

*Biozonation*

The charts shown in Text-figures 92–94 show the ranges of the principal species, plotted against a sequence of zones (it should be emphasized that the zones are used here in a strictly biostratigraphical context, and are not chronozones). The relationship between this biostratigraphy and the chronostratigraphical divisions of the British Coal Measures is shown in Text-figure 3. The zones are essentially those of Wagner (1984), except that an attempt has been made to simplify their nomenclature, following Cleal (1991a). In particular, Wagner's *Lyginopteris hoeninghausii/ Neuralethopteris schlehanii* and *Lonchopteris rugosa/Alethopteris urophylla* zones have been renamed the *L. hoeninghausii* and *Lonchopteris rugosa* zones respectively. In addition, the boundary between the *L. rugosa* and *Paripteris linguaefolia* zones has been redefined, to make it more easily identified.

Imposed on Wagner's set of zones are a number of subzonal divisions, based on the scheme proposed by Cleal (1991a). The subzones were based largely on distributional evidence derived from the European Paralic Coal Basin and so should be applicable throughout Britain, as well as probably in the nearby coalfields in northern France, Belgium, The Netherlands and northern Germany. However, no claim is made as to their significance in a wider context, which is why they are established as subzones rather than zones.

*1. Lyginopteris hoeninghausii Zone.* This coincides with the Langsettian Stage in Britain. The base of the zone can be recognized by the lowest

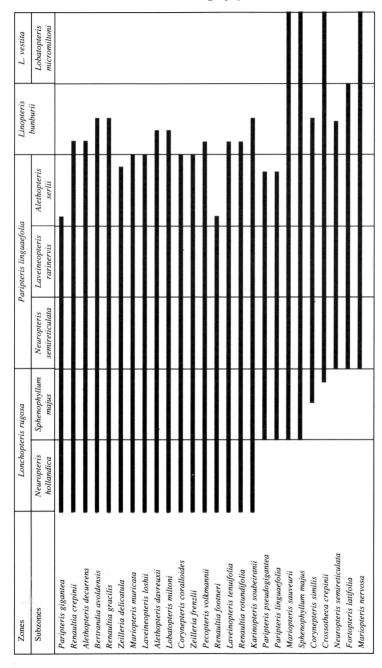

TEXT-FIG. 93. Main stratigraphical ranges of some of the key plant fossil species in the Duckmantian to the lower Westphalian D.

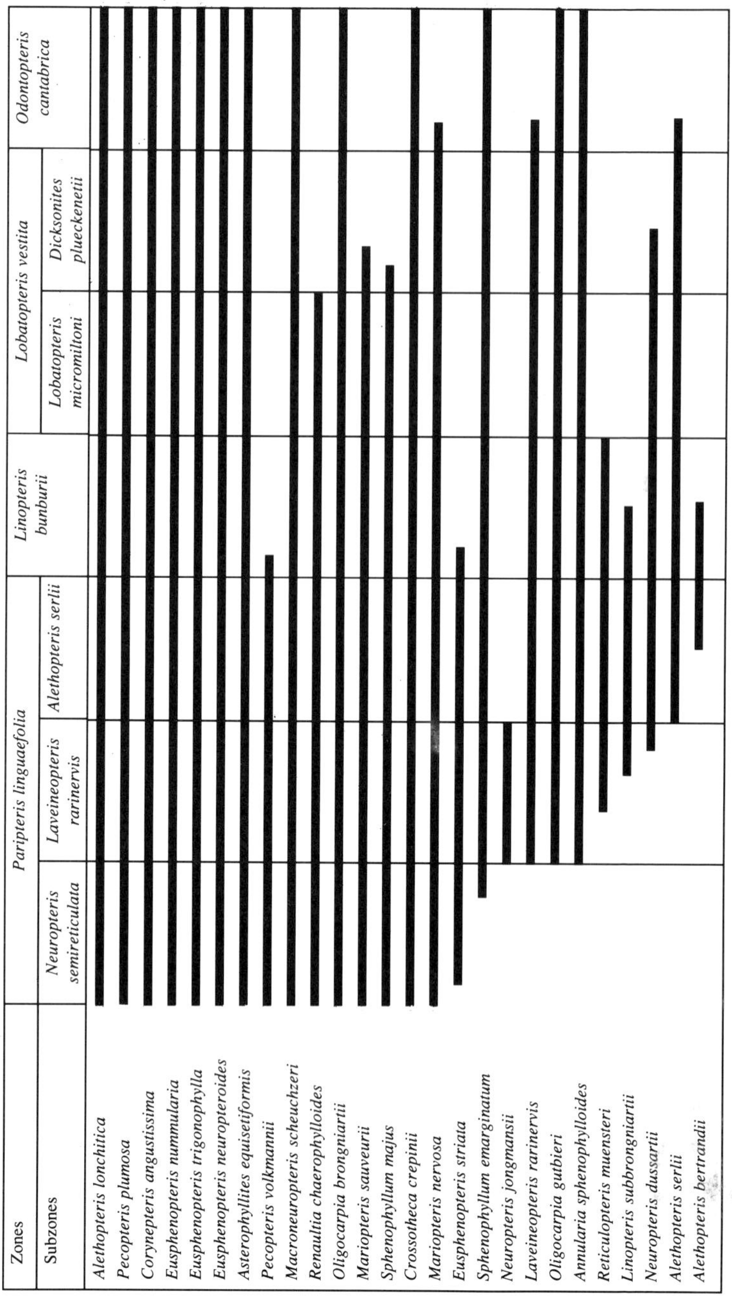

# British Coal Measures

| Zones | Paripteris linguaefolia | | | Linopteris bunburii | Lobatopteris vestita | | Odontopteris cantabrica |
|---|---|---|---|---|---|---|---|
| Subzones | Neuropteris semireticulata | Laveineopteris rarinervis | Alethopteris serlii | | Lobatopteris micromiltoni | Dicksonites plueckenetii | |

Alethopteris lonchitica
Pecopteris plumosa
Corynepteris angustissima
Eusphenopteris nummularia
Eusphenopteris trigonophylla
Eusphenopteris neuropteroides
Asterophyllites equisetiformis
Pecopteris volkmannii
Macroneuropteris scheuchzeri
Renaultia chaerophylloides
Oligocarpia brongniartii
Mariopteris sauveurii
Sphenophyllum majus
Crossotheca crepinii
Mariopteris nervosa
Eusphenopteris striata
Sphenophyllum emarginatum
Neuropteris jongmansii
Laveineopteris rarinervis
Oligocarpia gutbieri
Annularia sphenophylloides
Reticulopteris muensteri
Linopteris subbrongniartii
Neuropteris dussartii
Alethopteris serlii
Alethopteris bertrandii

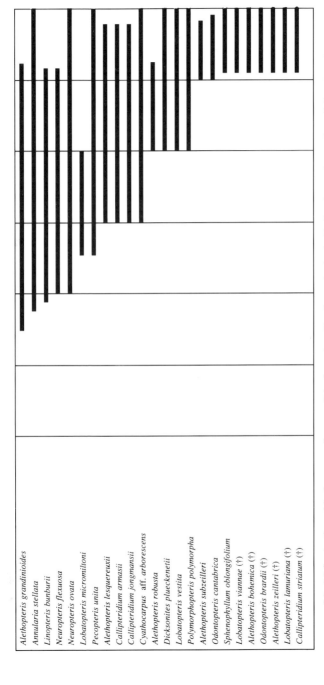

TEXT-FIG. 94. Main stratigraphical ranges of some of the key plant fossil species in the Bolsovian to lower Cantabrian. Species not yet reported from Britain marked with (†).

occurrences of *Neuralethopteris jongmansii, Lyginopteris hoeninghausii, Eusphenopteris nummularia, E. trigonophylla, E. grandis, Renaultia gracilis, Corynepteris angustissima* and *Asterophyllites charaeformis*. It represents a marked increase in species diversity compared with most Namurian assemblages. The range of *L. hoeninghausii* extends somewhat above the top of the zone as defined by Wagner, which is best recognized by the highest occurrences of *Neuralethopteris schlehanii* and *Karinopteris acuta*. In Britain, there is a marked difference in composition between assemblages found in the lower and upper parts of the zone.

1a. *Neuralethopteris jongmansii* Subzone. This coincides with the range of *N. jongmansii* in the lower Langsettian, and represents a transition from the typically Westphalian to the typically Namurian plant fossil assemblages. In Britain, the assemblages are typically dominated by neuralethopterids, lyginopterids and *Eusphenopteris hollandica*, and examples are found at Nant Llech near Abercraf, Powys (Dix 1933), on the coast near Saundersfoot, Pembrokeshire (Goode 1913), and several localities exposing the Paint Coal in Devon (Arber 1902).

1b. *Laveineopteris loshii* Subzone. The base of this subzone, which coincides with the base of Zone D of Dix (1934), is recognized by the lowest occurrences of *L. loshii, Neuropteris heterophylla, Margaritopteris conwayi, Neuralethopteris rectinervis, Alethopteris davreuxii, Eusphenopteris sauveurii, E. neuropteroides* and *Lobatopteris miltoni*. However, it is impossible to delineate the subzone by the range of a single species. It represents those *L. hoeninghausii* Zone assemblages in which lyginopterids, neuralethopterids and *E. hollandica* occur only rarely, and which are typically dominated by *L. loshii* (e.g. Thatto Heath, Lancashire–Cleal 1979, Cattybrook, Avon–Cleal and Thomas 1988). In many ways, it marks the lowest of the typical Coal Measures assemblages.

2. Lonchopteris rugosa *Zone*. This coincides approximately with the lower and middle Duckmantian in Britain. *L. rugosa* occurs both above and below the bounds of the zone, but the name has been retained for the interval for historical reasons. The base of the zone can be recognized by the lowest occurrences of *Renaultia rotundifolia, R. chaerophylloides, Oligocarpia brongniartii* and *Sphenophyllum myriophyllum*, and by the disappearance of *Karinopteris acuta* and *Neuralethopteris schlehanii*. The top of the zone, as defined by Wagner (1984) is difficult to recognize on purely palaeobotanical criteria and Cleal (1991a) proposed that it should be placed at a somewhat lower and easier to define horizon, which we have accepted here. There is also a marked change in the assemblages in the middle of the zone, and therefore two subzones are recognized.

2a. *Neuropteris hollandica* Subzone. The base of this interval is defined by the base of the *L. rugosa* Zone. The subzone is not delineated by the range of any particular species, but represents that interval between the base of the *L. rugosa* Zone and the appearance of those species marking the base of the overlying *Sphenophyllum majus* Subzone. There are few well documented assemblages from this subzone in Britain. Scott (1984) has described an assemblage from Swillington in West Yorkshire. The plant fossils associated with the Lower Haigh Moor Coal at Robin Hood Quarry, Leeds, may also belong to this subzone, although the illustrations provided by Walton (1933) are poor. The ironstone nodules from the Phoenix Brickworks at Crawcrook, Durham also probably belong here, although the full assemblage has not been fully documented (see Crookall 1969 for illustrations of some of the specimens).

2b. *Sphenophyllum majus* Subzone. The base of this subzone is recognized by the lowest occurrences of *Paripteris linguaefolia*, *P. pseudogigantea*, *Mariopteris sauveuri* and *Sphenophyllum majus*, and the disappearance of *Lyginopteris hoeninghausii* and *L. baeumleri*. The top of the subzone is defined by the base of the overlying *Paripteris linguaefolia* Zone. It includes some of the classic Westphalian assemblages in Britain, exemplified by that found associated with the Barnsley Thick Coal in South Yorkshire. There is no monographic treatment of this important assemblage, although many specimens have been described in the literature (e.g. Kidston 1895, 1896, 1923–1925; Crookall 1955–1976; Scott 1978; Shute and Cleal 1989; Cleal and Shute 1991). The Coseley ironstones found near Dudley have also yielded an important assemblage of this subzone (Kidston 1914; Arber 1916).

*3. Paripteris linguaefolia Zone.* It is proposed that the base of this zone is located at a lower and more easily recognizable horizon than suggested by Wagner (1984). The base of the zone is represented by the lowest occurrences of *Fortopteris latifolia*, *Mariopteris nervosa* and *Neuropteris semireticulata*, and the disappearance of *Neuropteris heterophylla*, *Alethopteris valida*, *Lonchopteris eschweileriana*, *Eusphenopteris obtusiloba* and *E. scribanii*. The top of the zone is defined by the base of the overlying *Linopteris bunburii* Zone, which coincides with the base of the Westphalian D Stage. Three subzones can be recognized within the zone.

3a. *Neuropteris semireticulata* Subzone. The base of this subzone coincides approximately with the base of the *N. semireticulata / L. tenuifolia* Subzone of Corsin and Corsin (1971). The subzone straddles the Duckmantian-Bolsovian stage boundary, which is difficult to locate using palaeobotanical data alone (Bouroz *et al.* 1969). This subzone is not particularly well represented in Britain, with lacustrine and marine deposits dominating many of the sequences, but an example from the West Cumberland Coalfield has recently been documented by Thomas and Cleal (1993).

3b. *Laveineopteris rarinervis* Subzone. The base of this subzone can be recognized by the lowest occurrences of *L. rarinervis, Oligocarpia gutbieri, Neuropteris jongmansii* and *Annularia sphenophylloides*, and the disappearance of *Neuropteris obliqua* and *Asterophyllites charaeformis*. The early examples of *Reticulopteris muensteri* also appear within the subzone. It marks an important change in the assemblages from an essentially middle Westphalian to an upper Westphalian aspect. Like the above subzone, the *L. rarinervis* Subzone has not been well documented in this country, one exception being from Nostell Priory Brickpit near Wakefield (Barker and Whittle 1944).

3c. *Alethopteris serlii* Subzone. There is a gradual change in the assemblages in the upper *P. linguaefolia* Zone, and the choice of subzonal boundary is somewhat arbitary. The lowest occurrence of *Alethopteris serlii* has been chosen here purely for historical reasons, since it coincides approximately with the boundary between Dix's (1934) zones F and G. Additional guides are the lowest occurrences of *Linopteris subbrongniartii* and *Neuropteris dussartii*, and the disappearance of *Eusphenopteris sauveurii*, which all occur a short distance below this level; and the lowest occurrence of *Alethopteris grandinioides*, and the disappearance of *Pecopteris pennaeformis* and *Zeilleria delicatula* a short distance above. This subzone is best represented in the South Wales Coalfield (Cleal 1978) although no individual assemblages have been described.

4. Linopteris bunburii *Zone*. This corresponds to the lower part of Zone H of Dix (1934). It was named the *Linopteris obliqua* Zone by Wagner (1984), but was renamed by Cleal (1984*b*) for taxonomic reasons. The base of the zone, which is an index to the Bolsovian–Westphalian D boundary, is marked by the lowest occurrence of *Neuropteris ovata* and *N. flexuosa*. As a guide to locating the boundary, Laveine (1977) pointed out that the lowest occurrences of *Annularia stellata* and *Linopteris bunburii*, and the disappearance of *Paripteris* spp. occur a short distance below this level (discussed further by Cleal 1978, 1984*a*, 1984*b*; Zodrow and Cleal 1985). The assemblages remain fairly uniform through this zone and no subzonal divisions are proposed. The best documented British assemblage of this subzone is from near Newgale on the Pembrokeshire coast (Cleal and Thomas 1991).

5. Lobatopteris vestita *Zone*. This marks an important change in the Coal Measures plant fossil assemblages, with the appearance of a number of species (e.g. *Cyathocarpus* spp, *Callipteridium* spp., *Polymorphopteris polymorpha, Dicksonites plueckenetii, Pecopteris unita*) which are normally more characteristic of Stephanian assemblages (Wagner 1984). Cleal (1978, 1984*b*) and Zodrow and Cleal (1985) found that the change occurred in two discrete steps, and it was proposed that two zones should

be established for this interval. To make the present classification conform with that of Wagner (1984), however, this two-fold transition will instead be recognized by a subzonal division within the *L. vestita* Zone.

5a. *Lobatopteris micromiltoni* Subzone. This is the same as the *L. micromiltoni* Zone of Cleal (1984*b*). The base of the subzone (and thus also of the *L. vestita* Zone) is placed at the lowest occurrences of *Alethopteris lesquereuxii* and *Cyathocarpus* ex group *arborescens*, and the lowest occurrences of abundant *Callipteridium jongmansii*, *Lobatopteris micromiltoni* and *Pecopteris unita*. Much of the strata of an age that would be expected to yield this subzone in Britain were removed by Variscan uplift and subsequent erosion. South Wales has yielded some examples from underground workings (Cleal 1978). The best documented assemblage from a surface outcrop is from Jockie's Syke in the Canonbie Coalfield (Kidston 1903).

5b. *Dicksonites plueckenetii* Subzone. This is the same as the *L. vestita* Zone of Cleal (1984*b*) (but not that of Wagner 1984). The base of the subzone is placed at the lowest occurrences of *Dicksonites plueckenetii*, *Polymorphopteris polymorpha* and *Lobatopteris vestita*. Britain has some of the classic assemblages of this subzone, perhaps most significantly those found in the Farrington and Radstock formations of the Somerset Coalfield (the so-called 'Radstock flora'—Kidston 1923–1925; Crookall 1955–1976; Thomas and Cleal 1994). It is also known from South Wales (Cleal 1978), Forest of Dean (Arber 1912; Cleal 1987) and the Severn Coalfield (Cleal 1986*b*).

6. Odontopteris cantabrica *Zone.* This is the highest zone recognized in the British Coal Measures (Cleal 1978). The base of the zone can be difficult to recognize (discussed by Zodrow and Cleal 1985), but is important as an index to the Westphalian–Stephanian series boundary. Many of the species used to define it do not occur in Britain, but the occurrence of *Odontopteris cantabrica* and *Alethopteris grandinioides* var. *subzeilleri* suggest that the *O. cantabrica* Zone is presence in the topmost Coal Measures of South Wales, the Forest of Dean and the Bristol area (Wagner and Spinner 1972; Cleal 1978, 1986*b*).

# GLOSSARY

The unnecessary use of technical jargon has become a bane of palaeobotany in recent years. Particularly within descriptions, however, technical terms can help maintain succinctness and, so long as their meanings are made clear, are a necessary evil. This section aims to define those terms that have been used in the keys, which may be unfamiliar to the non-specialist reader. In addition, a clarification of the terms used to described lycopsid leaf-scars (relevant to Key Group B) can be found in Text-figure 9.

Abaxial: Surface of leaf or pinnule facing the basal part of the plant. In effect, it is what is informally referred to as the 'lower' surface.

Acroscopic: The margin of a pinnule (q.v.) that faces the apex of the pinna (q.v.) on which it is borne.

Acuminate: Sharply pointed.

Adaxial: Surface of leaf or pinnule facing the apex of the plant. In effect, it is what is informally referred to as the 'upper' surface.

Anastomosed: Used to describe veins of a leaf or pinnule (q.v.) that form a mesh-like network, similar to that seen in most living flowering plants.

Aperture: Opening.

Apical pinnule: The pinnule borne at the apex of a pinna (q.v.), and which is usually of different shape to the pinnules borne laterally on the rachis.

Auricle: A small projection or 'ear' at the base of a pinnule (q.v.).

Basiscopic: The margin of a pinnule (q.v.) that faces the base of the pinna (q.v.) on which it is borne.

Biconvex: Barrel-shaped, with outwards-bulging lateral margins.

Bluntly acuminate: Pointed in the shape of a Gothic arch.

Bract: Part of the cone scale (q.v.) that protects the sporangia (q.v.).

Ciliated: Has fine hairs, resembling an eye-lash.

Clavate: Club-shaped; broad distal end, tapering proximally to a narrow base.

Cone scale: Leaf-like structure within some strobili (q.v.), which helps protect the sporangia (q.v.). Also sometimes known as a sporophyll.

## Glossary

| | |
|---|---|
| Confluent: | Adjacent pinnules or pinnule-lobes that are basally fused. |
| Conical: | Cone-shaped. |
| Contiguous: | Touching neighbours. |
| Cordate: | With two equal, rounded lobes; heart-shaped. |
| Cuneiform: | Inversely subtriangular, which is narrowest at point of attachment. |
| Decurrent: | Strongly curved near base. |
| Dehiscence: | Splitting |
| Dichotomous: | Forks into two equal branches. |
| Digitate: | With finger-like projections. |
| Drip-tip: | An elongate extension of a leaf or pinnule, thought to facilitate the run-off of rain-water. |
| Eccentric: | Off-centre. |
| Granulate: | Surface covered by fine, evenly-distributed projections or bumps. |
| Dentate: | Margin with tooth-like projections. |
| Distal: | Away from point of attachment. |
| Falcate: | Sickle-shaped. |
| Fimbriate: | A ragged margin. |
| Fusiform: | Elongately spindle-shaped, swollen in the middle and tapering towards either end. |
| Incised: | Cut into. |
| Inflexed: | Bent inwards. |
| Internodes: | The section of a stem between adjacent nodes (q.v.). |
| Isodiametric: | Equally long as wide. |
| Keel: | Ridge, similar in shape to that running along the bottom of a ship. |
| Laciniate: | Slashed into deep, narrow divisions, with tapering, pointed incisions. |
| Lanceolate: | Elongate with curved sides, at its broadest in the middle. |
| Lateral veins: | Veins that are emitted from the midvein (q.v.) and extend towards the margin of the pinnule (q.v.). |
| Lax: | Of flaccid, non-rigid appearance. |
| Limb: | When referring to a pinnule (q.v.), it is the laminate structure, including both the veins and the inter-vein tissue. |
| Linear: | Elongate and parallel-sided. |
| Linguaeform: | Tongue-shaped. |
| Micropyle: | Opening through which the pollen enters to fertilize an ovule to produce a seed. |
| Midvein: | Main vein which runs along the long-axis of a pinnule (q.v.). |

| | |
|---|---|
| Nervation: | The pattern of veining in a leaf or pinnule (q.v.). In the latter, it usually consists of a midvein (q.v.) and lateral veins (q.v.). |
| Nervation density: | The number of veins per cm that meet the pinnule (q.v.) margin. |
| Nodes: | The position on a stem where a leaf or branch is attached. In the horsetails, it refers to the transverse marks across a stem where the characteristic whorls of leaves and/or branches are attached (Text-figs 23–25). |
| Oblanceolate: | Similar to lanceolate (q.v.), but broadest above the middle. |
| Obtuse: | Blunt. |
| Ovate: | Near-oval or egg-shaped. |
| Pedicel: | Stalk attaching a sporangium (q.v.) to an axis. |
| Petiole: | Of a leaf or frond, it is the stalk that connects the laminate structure to the stem. Of a pinnule (q.v.), it is the stalk that connects the limb (q.v.) to the rachis (q.v.). |
| Pinna(e): | A division of a compound leaf or frond, consisting of a rachis (q.v.) which bears pinnules (q.v.) or lower-order pinnae. |
| Pinnatifid: | Lobed pinnule, representing an intermediate stage between an entire-margined pinnule (q.v.) and a pinna (q.v.). |
| Pinnule: | The ultimate division of a compound leaf or frond, which is clearly laminate. |
| Proximal: | Near to place of attachment. |
| Pseuodoanastomosed: | Used to describe veins in pinnules (q.v.), which have become so flexuous that neighbouring veins touch. They thus resemble anastomosed (q.v.) veins, except that the touching veins retain their individuality, and do not become fully fused together. |
| Punctae: | Fine, dot-like markings. |
| Pyriform: | Pear-shaped. |
| R1, R2, R3: | See Rachis. |
| Rachis: | 'Axis' in a compound leaf or frond, which bears pinnae (q.v.) or pinnules (q.v.). The main 'axis' of the frond is known as a primary (R1) rachis. This bears secondary (R2) rachises, which in turn bear tertiary (R3) rachises. |
| Radiating: | Spreads out from a point of origin. |
| Reniform: | Kidney-shaped. |

## Glossary

| | |
|---|---|
| Reticulation: | Mesh-work. |
| Rhomboidal: | Diamond-shaped, but sometimes with one half longer than the other. |
| Sessile: | Directly attached, without a stalk or petiole (q.v.). |
| Sori: | Discrete clusters of sporangia (q.v.) found in some ferns. |
| Spatulate: | Broad at upper end, tapering to a narrow base, i.e. spoon-shaped. |
| Sporangiophore: | A structure bearing one or more sporangia (q.v.). It may be simply a short stalk, or part of a cone scale (q.v.). |
| Sporangium: | Small, sack-like body that produces spores. In most ferns, they are borne on the underside (abaxial surface) of the pinnules. In the lycopsids (club mosses) and sphenopsids (horsetails) they are borne in strobili (q.v.). In seed-bearing plants, they are borne in clusters of a variety of forms. |
| Sporophyll: | See cone scale. |
| Striations: | Fine, linear markings. |
| Strobilus: | A reproductive structure consisting of sporangia or clusters of sporangia (q.v.) arranged spirally or in whorls around a stem-like axis. The sporangia/sporangial clusters may be attached to cone scales (q.v.), or directly to the axis between cone scales. The whole strobilus often resembles a conifer cone in shape, and is sometimes referred to informally as a 'cone'. |
| Sub-: | Used as a prefix to some of the descriptive terms, implying that the object is tending towards that shape (e.g. subtriangular—tending towards a triangular shape, but perhaps with somewhat curved margins). |
| Tubercle: | A wart-like projection or bump. |

# REFERENCES

ABBOTT, M. L. 1958. The American species of *Asterophyllites, Annularia*, and *Sphenophyllum. Bulletins of American Paleontology*, **38**, 289–390.

AMEROM, H. W. J. VAN 1975. Die eusphenopteridischen Pteridophyllen aus der Sammlung des Geologischen Bureaus in Heerlen, unter besonderer Berücksichtigung ihrer Stratigraphie bezüglich des Südlimburger Kohlenreviers. *Mededelingen Rijks Geologische Dienst, Serie C*, **3–1–7**, 1–201.

ANDREWS, H. N. 1980. *The fossil hunters. In search of ancient plants.* Cornell University Press, Ithaca and London, 421 pp.

ARBER, E. A. N. 1902. The fossil flora of the Culm Measures of north-west Devon, and the palæobotanical evidence with regard to the age of the beds. *Philosophical Transactions of the Royal Society of London, Series B*, **197**, 291–325.

—— 1912. On the fossil flora of the Forest of Dean Coalfield (Gloucestershire), and the relationships of the coalfields of the west of England and South Wales. *Philosophical Transactions of the Royal Society of London, Series B*, **202**, 233–280.

—— 1916. On the fossil floras of the Coal Measures of South Staffordshire. *Philosophical Transactions of the Royal Society of London, Series B*, **208**, 127–155.

AUSTIN, R., BISHOP, M. J., CLEAL, C. J., COPP, C. J. T., HARLEY, M. J., SYMES, R. F. and WOOD, C. 1985. *New sites for old. A student's guide to the geology of the east Mendips*. Nature Conservancy Council, Peterborough, 192 pp.

BARKER, W. R. and WHITTLE, W. L. 1944. The Coal Measure strata of Nostell, near Wakefield. *Proceedings of the Yorkshire Geological Society*, **25**, 175–189.

BARTHEL, M. 1961. Der Epidermisbau einiger oberkarbonischer Pteridospermen. *Geologie*, **10**, 828–849.

—— 1962. Epidermisuntersuchungen an einigen inkohlten Pteridospermenblättern des Oberkarbons und Perms. *Geologie*, **11**, 1–140.

—— 1968. '*Pecopteris' feminaeformis* (Schlotheim) Sterzel und '*Araucarites' spiciformis* Andrae in Germar—Coenopteriden des Stephans und unteren Perms. *Paläontologische Abhandlungen*, **2**, 727–742.

BASSETT, D. A. and BASSETT, M. G. (eds). 1971. *Geological excursions in South Wales and the Forest of Dean*. Geologists' Association South Wales Group, Cardiff, 267 pp.

BASSETT, M. G. (ed.) 1982. *Geological excursions in Dyfed, south-west Wales*. National Museum of Wales, Cardiff, 327 pp.

BATEMAN, R. M. 1991. Palaeoecology. 34–116. *In* CLEAL, C. J. (ed.). *Plant fossils in geological investigation: the Palaeozoic*. Ellis Horwood, Chichester, 233 pp.

BATENBURG, L. H. 1977. The *Sphenophyllum* species in the Carboniferous flora of Holz (Westphalian D, Saar Basin, Germany). *Review of Palaeobotany and Palynology*, **24**, 69–99.

—— 1981. Vegetative anatomy and ecology of *Sphenophyllum zwickaviense*,

*S. emarginatum*, and other "compression species" of *Sphenophyllum*. *Review of Palaeobotany and Palynology*, **32**, 275–313.

BERTRAND, P. 1914. Les zones végétales du terrain houiller du Nord de la France. Leur extension verticale par rapport aux horizons marins. *Annales de la Société Géologique du Nord*, **43**, 208–254.

BESLY, B. M. 1988. Palaeogeographic implications of late Westphalian to early Permian red-beds, Central England. 200–221. *In* BESLY, B. M. and KELLING, G. (eds). *Sedimentation in a synorogenic basin complex. The Upper Carboniferous of northwest Europe*. Blackie, Glasgow, and Chapman and Hall, New York, 276 pp.

BOERSMA, M. 1972. The heterogeneity of the form genus *Mariopteris* Zeiller. Thesis, University of Utrecht.

BOULTER, M. C. 1968. On a species of compressed lycopod sporophyll from the Upper Coal Measures of Somerset. *Palaeontology*, **11**, 445–457.

—— CHALONER, W. G. and HOLMES, P. L. 1991. The IOP Plant Fossil Record: are fossil plants a special case? 231–242. *In* HAWKSWORTH, D. L. (ed.). Improving the stability of names: needs and options. *Regnum Vegetabile*, **123**, 1–358.

BOUREAU, E. 1964. *Traité de paléobotanique III. Sphenophyta, Noeggerathiophyta*. Masson et Cie, Paris, 544 pp.

—— and DOUBINGER, J. 1975. *Traité de paléobotanique IV (2). Pteridophylla. Première partie*. Masson et Cie, Paris, 768 pp.

BOUROZ, A., CHALARD, J., CORSIN, P. and LAVEINE, J.-P. 1969. Le stratotype du Westphalien C dans le bassin houiller du Nord et du Pas-de-Calais: limites et contenue paléontologique. *Compte Rendu—6ᵉ Congrès International de Stratigraphie et de Géologie du Carbonifère (Sheffield, 1967)*, **1**, 99–105.

—— EINOR, O. L., GORDON, M., MEYEN, S. V. and WAGNER, R. H. 1978. Proposals for an international stratigraphic classification of the Carboniferous. *Compte Rendu—8ᵉ Congrès de Stratigraphie et de Géologie du Carbonifère (Moscow, 1975)*, **1**, 36–69.

BRACK-HANES, S. D. and THOMAS, B. A. 1983. A re-examination of *Lepidostrobus* Brongniart. *Botanical Journal of the Linnean Society*, **86**, 125–133.

BRENCKLE, P. L. and MANGER, W. L. (eds). 1991. International correlation and division of the Carboniferous System. Contributions from the Carboniferous Subcommission Meeting, Provo, Utah, September 1989. *Courier Forschungsinstitut Senckenberg*, **130**, 1–350.

BRIGGS, D. E. G. and CROWTHER, P. R. (eds). 1990. *Palaeobiology. A synthesis*. Blackwell, Oxford, 583 pp.

BROADHURST, F. M. 1988. Seasons and tides in the Westphalian. 264–272. *In* BESLY, B. M. and KELLING, G. (eds). *Sedimentation in a synorogenic basin complex. The Upper Carboniferous of northwest Europe*. Blackie, Glasgow, and Chapman and Hall, New York, 276 pp.

—— and FRANCE, A. A. 1986. Time represented by coal seams in the Coal Measures of England. *International Journal of Coal Geology*, **6**, 43–54.

—— SIMPSON, I. M. and HARDY, P. G. 1980. Seasonal sedimentation in the Upper Carboniferous of England. *Journal of Geology*, **88**, 639–651.

BRONGNIART, A. 1828–1838. *Histoire des végétaux fossiles*. 2 Volumes. Paris, 560 pp.

BROUSMICHE, C. 1983. *Les Fougères sphénoptéridiennes du Bassin Houiller Sarro-*

*Lorraine.* Société Géologique du Nord, Lille (Publication No. 10), 480pp.
BUISINE, M. 1961. Les Aléthoptéridées du Nord de la France. *Étude Géologiques pour l'Atlas de Topographie Souterraine*, **1**(4), 1–317.
CALVER, M. A. 1968. Distribution of Westphalian marine faunas in northern England and adjoining areas. *Proceedings of the Yorkshire Geological Society*, **37**, 1–72.
CARROLL, R. L. 1984. Problems in the use of terrestrial vertebrates for zoning the Carboniferous. *Compte Rendu—9ᵉ Congrès de Stratigraphie et de Géologie du Carbonifère (Washington and Urbana, 1979)*, **2**, 135–147.
CHALONER W. G. 1953. On the megaspores of four species of *Lepidostrobus*. *Annals of Botany*, **17**, 263–293.
—— 1958. The Carboniferous upland flora. *Geological Magazine*, **95**, 261–262.
—— 1967. Lycophyta. 435–802. *In* BOUREAU, E., JOVEST-AST, S., HØEG, O. A. and CHALONER, W. G. *Traité de Paléobotanique II*. Masson et Cie, Paris, 845pp.
—— 1986. Reassembling the whole plant, and naming it. 67–78. *In* SPICER, R. A. and THOMAS, B. A. (eds). *Systematic and taxonomic approaches in palaeobotany*. Systematics Association, London (Special Volume No. 31), 318 pp.
—— and COLLINSON, M. E. 1975a. Application of SEM to a sigillarian impression fossil. *Review of Palaeobotany and Palynology*, **20**, 85–101.
———— 1975b. An illustrated key to the commoner British Upper Carboniferous plant compression fossils. *Proceedings of the Geologists' Association*, **86**, 1–44.
CLEAL, C. J. 1978. Floral biostratigraphy of the upper Silesian Pennant Measures of South Wales. *Geological Journal*, **13**, 165–194.
—— 1979. The Ravenhead Collection of fossil plants. *Amateur Geologist*, **9**(11), 12–23.
—— 1984a. The recognition of the base of the Westphalian D Stage in Britain. *Geological Magazine*, **121**, 125–129.
—— 1984b. The Westphalian D floral biostratigraphy of Saarland (Fed. Rep. Germany) and a comparison with that of South Wales. *Geological Journal*, **19**, 327–351.
—— 1985. The Cyclopteridaceae (Medullosales, Pteridospermopsida) of the Carboniferous of Saarland, Federal Republic of Germany. Unpublished PhD thesis, University of Sheffield.
—— 1986a. Identifying plant fragments. 53–65. *In* SPICER, R. A. and THOMAS, B. A. (eds). *Systematic and taxonomic approaches in palaeobotany*. Systematics Association, London (Special Volume No. 31), 318 pp.
—— 1986b. Fossil plants of the Severn Coalfield and their biostratigraphical significance. *Geological Magazine*, **123**, 553–568.
—— 1987. Macrofloral biostratigraphy of the Newent Coalfield, Gloucestershire. *Geological Journal*, **22**, 207–217.
—— 1991a. Carboniferous and Permian biostratigraphy. 182–215. *In* CLEAL, C. J. (ed.). *Plant fossils in geological investigation: the Palaeozoic*. Ellis Horwood, Chichester, 233 pp.
—— 1991b. The age of the base of the Forest of Dean Coal Measures: fact and fancy. *Proceedings of the Geologists' Association*, **102**, 261–264.
—— and SHUTE, C. H. 1991. The Carboniferous pteridosperm frond *Neuropteris*

## References

*heterophylla* (Brongniart) Sternberg. *Bulletin of the British Museum (Natural History), Geology Series*, **46**, 153–174.
—— —— 1992. Epidermal features of some Carboniferous neuropteroid fronds. *Revew of Palaeobotany and Palynology*, **71**, 191–206.
—— —— and ZODROW, E. L. 1989. A revised taxonomy for Palaeozoic neuropteroid foliage. *Taxon*, **39**, 486–492.
—— and THOMAS, B. A. 1988. The Westphalian fossil floras from the Cattybrook Claypit, Avon (Great Britain). *Geobios*, **21**, 409–433.
—— —— 1991. Carboniferous and Permian palaeogeography. 154–181. *In* CLEAL, C. J. (ed.). *Plant fossils in geological investigation: the Palaeozoic*. Ellis Horwood, Chichester, 233 pp.
—— —— 1992. Lower Westphalian D fossil plants from the Nolton-Newgale Coalfield, Dyfed (Great Britain). *Geobios*, **25**, 315–322.
—— —— in press. *Geological Conservation Review. Palaeobotany (1)*. Chapman and Hall, London.
—— and ZODROW, E. L. 1989. Epidermal structure of some medullosan *Neuropteris* foliage from the middle and upper Carboniferous of Canada and Germany. *Palaeontology*, **32**, 837–882.
COPE, J. C. W., GUION, P. D., SEVASTOPULO, G. D. and SWAN, A. R. H. 1992. Carboniferous. 67–86. *In* COPE, J. C. W., INGHAM, K. J. and RAWSON, P. F. (eds). *Atlas of palaeogeography and lithofacies*. The Geological Society, London, 153 pp.
COPE, M. J. and CHALONER, W. G. 1985. Wildfire, an interaction of biological and physical processes. 257–277. *In* TIFFNEY, B. H. (ed.). *Geological factors and the evolution of plants*. Yale University Press, Hartford CT, 294 pp.
CORSIN, P. and CORSIN, P. 1971. Zonation biostratigraphique du Houiller des bassins du Nord-Pas-de-Calais et de Lorraine. *Compte Rendu Hebdomaire des Séances de l'Académie des Sciences, Paris, Série D*, **273**, 783–788.
COX, B., SAVAGE, R. J. G., GARDINER, B. and DIXON, D. 1988. *MacMillan illustrated encyclopedia of dinosaurs and prehistoric animals*. Guild Publishing, London, 312 pp.
CRAIG, G. Y. (ed.). 1991. *The geology of Scotland (3rd edition)*. Geological Society, London, 612 pp.
CRANE, P. R. 1985. Phylogenetic analysis of seed plants and the origin of angiosperms. *Annals of the Missouri Botanical Gardens*, **72**, 716–793.
CROOKALL, R. 1955–1976. Fossil plants of the Carboniferous rocks of Great Britain. [Second section]. *Memoirs of the Geological Survey of Great Britain, Palaeontology*, **4**, Part 1 (1955), 1–84, Part 2 (1957), 85–216, Part 3 (1964), 217–354, Part 4 (1966), 355–572, Part 5 (1969), 573–792, Part 6 (1970), 793–840, Part 7 (1976), 841–1008.
DALINVAL, A. 1960. Les *Pecopteris* du bassin houiller du Nord de la France. *Étude Géologiques pour l'Atlas de Topographie Souterraine*, **1**(3), 1–222.
DANZÉ-CORSIN, P. 1953. Contribution à l'étude des Marioptéridées. Les *Mariopteris* du Nord de la France. *Étude Géologiques pour l'Atlas de Topographie Souterraine*, **1**(1), 1–269.
DARRAH, W. C. 1969. *A critical review of the Upper Pennsylvanian floras of eastern United States with notes on the Mazon Creek flora of Illinois*. W. C. Darrah, Gettysburg, 220 pp.

DIMICHELE, W. A. and ARONSON, R. B. 1992. The Pennsylvanian-Permian vegetational transition: a terrestrial analogue to the onshore-offshore hypothesis. *Evolution*, **46**, 807–824.

—— and PHILLIPS, T. L. 1985. Arborescent lycopod reproduction and paleoecology in a coal-swamp environment of late Middle Pennsylvanian age (Herrin Coal, Illinois, U.S.A.). *Review of Palaeobotany and Palynology*, **44**, 1–26.

—— —— and PEPPERS, R. A. 1985. The influence of climate and depositional environment on the distribution and evolution of Pennsylvanian coal-swamp plants. 223–256. *In* TIFFNEY, B. H. (ed.). *Geological factors and the evolution of plants*. University Press, Yale, 294 pp.

DIX, E. 1933. The succession of fossil plants in the Millstone Grit and the lower portion of the Coal Measures of the South Wales Coalfield (near Swansea) and a comparison with that of other areas. *Palaeontographica, Abteilung B*, **78**, 157–202.

—— 1934. The sequence of floras in the Upper Carboniferous, with special reference to South Wales. *Transactions of the Royal Society of Edinburgh*, **57**, 789–838.

DUFF, P. McL. D. and SMITH A. J. (eds). 1992. *Geology of England and Wales*. Geological Society, London, 651 pp.

DURDEN, C. J. 1984. Carboniferous and Permian entomology of western North America. *Compte Rendu—9ᵉ Congrès de Stratigraphie et de Géologie du Carbonifère (Washington and Urbana, 1979)*, **2**, 81–89.

ELDREDGE, N. and GOULD, S. J. 1972. Punctuated equilibria: an alternative to phyletic gradualism. 85–115. *In* SCHOPF, T. J. M. (ed.). *Models in paleobiology*. Freeman, San Francisco, vi + 250 pp.

ELLIOTT, R. E. 1968. Facies, sedimentation successions and cyclothems in Productive Coal Measures in the East Midlands, Great Britain. *Mercian Geologist*, **2**, 351–372.

—— 1969. Deltaic processes and episodes: the interpretation of productive Coal Measures occurring in the East Midlands. *Mercian Geologist*, **3**, 111–134.

ENGEL, B. A. 1989. S.C.C.S. ballot results. *Newsletter on Carboniferous Stratigraphy*, **7**, 6–8.

FIELDING, C. R. 1984*a*. Upper delta plain lacustrine and fluviolacustrine facies from the Westphalian of the Durham coalfield, NE England. *Sedimentology*, **31**, 547–567.

—— 1984*b*. A coal depositional model for the Durham Coal Measures of NE England. *Journal of the Geological Society, London*, **141**, 919–931.

—— 1986. Fluvial channel and overbank deposits from the Westphalian of the Durham coalfield, NE England. *Sedimentology*, **33**, 119–140.

—— 1987. Lower delta plain interdistributary deposits—an example from the Westphalian of the Lancashire Coalfield, northwest England. *Geological Journal*, **22**, 151–162.

FISSUNENKO, O. P. and LAVEINE, J.-P. 1984. Comparaison entre la distribution des principales espèces-guides végétales du Carbonifère moyen dans le bassin du Donetz (URSS) et les bassins du Nord-Pas-de-Calais et de Lorraine (France). *Compte Rendu—9ᵉ Congrès de Stratigraphie et de Géologie du Carbonifère (Washington and Urbana, 1979)*, **1**, 95–100.

GASTALDO, R. A. 1981. Taxonomic considerations for Carboniferous coalified

compression equisetalean strobili. *American Journal of Botany*, **68**, 1319–1324.
—— 1985. Upper Carboniferous paleoecological reconstructions: observations and reconsiderations. *Compte Rendu — 10ᵉ Congrès International de Stratigraphie et de Géologie du Carbonifère (Madrid, 1983)*, **2**, 281–296.
—— 1987. Confirmation of Carboniferous clastic swamp communities. *Nature, London*, **326**, 869–871.
—— DOUGLAS, D. P. and McCARROLL, S. M. 1987. Origin, characteristics, and provenance of plant macrodetritus in a Holocene crevasse splay, Mobile Delta, Alabama. *Palaios*, **2**, 229–240.
—— GIBSON, M. A. and GRAY, T. D. 1989. An Appalachian-sourced deltaic sequence, north-eastern Alabama, U.S.A.: biofacies-lithofacies relationships and interpreted community patterns. *International Journal of Coal Geology*, **12**, 225–257.
GOODE, R. H. 1913. On the fossil flora of the Pembrokeshire portion of the South Wales coalfield. *Quarterly Journal of the Geological Society, London*, **69**, 252–276.
GRAND'EURY, F.-C. 1877. Flore carbonifère du Département de la Loire et du centre de la France. *Mémoires de l'Académie des Sciences de l'Institut de France*, **24**, 1–624.
GUION, P. D. 1984. Crevasse splay deposits and roof-rock quality in the Threequarters Seam (Carboniferous) in the East Midlands Coalfield, U.K. *Special Publications, International Association of Sedimentologists*, **7**, 291–308.
—— 1987*a*. The influence of a palaeochannel on seam thickness in the Coal Measures of Derbyshire, England. *International Journal of Coal Geology*, **7**, 269–299.
—— 1987*b*. Palaeochannels in mine workings in the High Hazels Coal (Westphalian B), Nottinghamshire Coalfield, England. *Journal of the Geological Society, London*, **144**, 471–488.
—— and FIELDING, C. R. 1988. Westphalian A and B sedimentation in the Pennine Basin, UK. 153–177. *In* BESLY, B. M. and KELLING, G. (eds). *Sedimentation in a synorogenic basin complex. The Upper Carboniferous of northwest Europe* Blackie, Glasgow, and Chapman and Hall, New York, 276 pp.
HALLE, T. G. 1933. The structure of certain fossil spore-bearing organs, believed to belong to pteridosperms. *Kungliga Svenska Vetenskapakademiens Handlingar, Series 3*, **12**, 1–103.
HASZELDINE, R. S. 1983*a*. Descending tabular cross-bed sets and bounding surfaces from a fluvial channel in the Upper Carboniferous coalfield of north-east England. *Special Publications, International Association of Sedimentologists*, **6**, 449–456.
—— 1983*b*. Fluvial bars reconstructed from a deep, straight channel, Upper Carboniferous coalfield of northeast England. *Journal of Sedimentary Petrology*, **53**, 1233–1247.
—— 1984. Muddy deltas in freshwater lakes, and tectonism in the Upper Carboniferous coalfield of NE England. *Sedimentology*, **31**, 811–822.
HAVLENA, V. 1971. Die zeitgleichen Floren des europäischen Oberkarbons und die mesophile Flora des Ostrau-Karwiner Steinkohlenreviers. *Review of Palaeobotany and Palynology*, **12**, 254–270.

HEDBERG, H. D. (ed.). 1976. *International stratigraphic guide. A guide to stratigraphic classification, terminology and procedure*. Wiley, London, 200pp.

HEIDE, S. VAN DER 1952. Troisième Congrès de Stratigraphie et Géologie du Carbonifère. *Compte Rendu—Troisième Congrès pour l'Avancement des Études de Stratigraphie et de Géologie du Carbonifère (Heerlen, 1951)*, **1**, vii–xxiii.

HIRMER, M. 1927. *Handbuch der Paläobotanik*. R. Oldenburg, Munich, 708 pp.

HOPPING, C. A. 1956. A note on the leaf cushion of a species of Palaeozoic arborescent lycopod (= *Sublepidophloios ventricosus* sp. nov.). *Proceedings of the Royal Society of Edinburgh, Series B*, **66**, 1–9.

JARZEMBOWSKI, E. A. 1987. The occurrence and diversity of Coal Measure insects. *Journal of the Geological Society, London*, **144**, 507–511.

—— 1989. Writhlington Geological Nature Reserve. *Proceedings of the Geologists' Association*, **100**, 219–234.

JONGMANS, W. J. 1928. Discussion générale. *Compte Rendu—Congrès pour l'Avancement des Études de Stratigraphie Carbonifère (Heerlen 1927)*, xxii–xlviii.

—— and GOTHAN, W. 1937. Betrachtung über die Ergebnisse des zweiten Kongresses für Karbonstratigraphie. *Compte Rendu—Deuxième Congrès pour l'Avancement des Études de Stratigraphie Carbonifère (Heerlen 1935)*, **1**, 1–40.

JOSTEN, K.-H. 1962. *Neuropteris semireticulata*, eine neue Art als Bindeglied zwischen den Gattungen *Neuropteris* und *Reticulopteris*. *Paläontologische Zeitschrift*, **36**, 33–45.

KELLING, G. 1974. Upper Carboniferous sedimentation in South Wales. 185–224. *In* OWEN, T. R. (ed.). *The Upper Palaeozoic and post-Palaeozoic rocks of Wales*. University of Wales Press, Cardiff, 426 pp.

KIDSTON, R. 1895. On the fossil flora of the Yorkshire Coal Field. (First paper). *Transactions of the Royal Society of Edinburgh*, **38**, 203–223.

—— 1896. On the fossil flora of the Yorkshire Coal Field. (Second paper). *Tranactions of the Royal Society of Edinburgh*, **39**, 33–62.

—— 1903. The fossil plants of the Carboniferous rocks of Canonbie, Dumfriesshire, and of parts of Cumberland and Northumberland. *Tranactions of the Royal Society of Edinburgh*, **40**, 741–833.

—— 1914. On the fossil flora of the Staffordshire coal fields. Part III. The fossil flora of the Westphalian Series of the South Staffordshire Coal Field. *Tranactions of the Royal Society of Edinburgh*, **50**, 73–190.

—— 1923–1925. Fossil plants of the Carboniferous rocks of Great Britain. *Memoirs of the Geological Survey of Great Britain, Palaeontology*, **2**, Parts 1–4 (1923), 1–376, Part 5 (1924), 377–522, Part 6 (1925), 523–670.

KNIGHT, J. A. 1974. The Stephanian A-B flora and stratigraphy of the Sabero Coalfield (León, N.W. Spain). *Compte Rendu—7$^{\underline{e}}$ Congrès de Stratigraphie et de Géologie du Carbonifère (Krefeld, 1971)*, **3**, 283–315.

KRAUSSE, H. F. and PILGER, A. 1977. Geotectonic and time relationship of Varsican foredeeps in central and western Europe (Subvariscan foredeep and Cantabro-Pyreneean foredeep). *Colloque International du Centre National de la Recherche Scientifique*, **243**, 547–557.

LAVEINE, J.-P. 1967. Les Neuroptéridées du Nord de la France. *Étude Géologiques pour l'Atlas de Topographie Souterraine*, **1**(5), 344 pp.

—— 1977. Report on the Westphalian D. 71–87. *In* HOLUB, V. M. and WAGNER, R. H. (eds). *Symposium on Carboniferous stratigraphy*. Geological Survey, Prague, 468 pp.

—— COQUEL, R. and LOBOZIAK, S. 1977. Phylogénie générale des Calliptéridiacées (Pteridospermopsida). *Geobios*, **10**, 757–847.

—— LEMOIGNE, Y., LI, X., WU, X., ZHANG, S., ZHAO, X., ZHU, W. and ZHU, J. 1987. Paléogéographie de la Chine au Carbonifère à la lumière des données paléobotaniques, par comparaison avec les assemblages carbonifères d'Europe occidentale. *Compte Rendu Hebdomaire des Séances de l'Académie des Sciences, Paris, Série D*, **304**, 391–394.

LEARY, R. L. 1975. Early Pennsylvanian paleogeography of an upland area, western Illinois, USA. *Bulletin Société Belge de Géologie*, **84**, 19–31.

—— 1977. Paleobotanical and geological interpretations of paleoenvironments of the Eastern Interior Basin. 157–164. *In* ROMANS, R. C. (ed.). *Geobotany*. Plenum Publishing Co., New York, viii + 308 pp.

—— and THOMAS, B. A. 1989. *Lepidodendron aculeatum* with attached foliage: evidence of stem morphology and fossilization processes. *American Journal of Botany*, **76**, 282–288.

LECKWIJCK, W. P. VAN 1960. Report of the Subcommission on Carboniferous Stratigraphy. *Compte Rendu—Quatrième Congrès pour l'Avancement des Études de Stratigraphie et de Géologie du Carbonifère (Heerlen, 1958)*, **1**, xxiv-xxvi.

LEEDER, M. R. 1988. Recent developments in Carboniferous geology: a critical review with implications for the British Isles and N.W. Europe. *Proceedings of the Geologists' Association*, **99**, 73–100.

LYONS, P. C. and DARRAH, W. C. 1989. Earliest conifers of North America: upland and/or paleoclimatic indicators? *Palaios*, **4**, 480–486.

MAMAY, S. H. 1992. *Sphenopteridium* and *Telangiopsis* in a *Diplopteridium*-like association from the Virgilian (Upper Pennsylvanian) of New Mexico. *American Journal of Botany*, **79**, 1092–1101.

—— and BATEMAN, R. M. 1991. *Archaeocalamites lazarii*, sp. nov.: the range of Archaeocalamitaceae extended from the lowermost Pennsylvanian to the mid-Lower Permian. *American Journal of Botany*, **78**, 489–496.

MANGER, W. L. 1985. Minutes of the meeting of west European titular members of S.C.C.S., September 16, 1983. *Compte Rendu—10ᵉ Congrès International de Stratigraphie et de Géologie du Carbonifère (Madrid, 1983)*, **4**, 459–460.

MAPES, G. and ROTHWELL, G. W. 1991. Structure and relationships of primitive conifers. *Neues Jahrbuch für Geologie und Paläontologie, Abhandlungen*, **183**, 269–287.

MILLAY, M. A. and TAYLOR, T. N. 1979. Paleozoic seed fern pollen organs. *Botanical Review*, **45**, 301–375.

MILNER, A. R. 1980. The tetrapod assemblage from Nýřany, Czechoslovakia. 439–496. *In* PANCHEN, A. L. (ed.). *The terrestrial environment and the origin of land vertebrates*. Systematics Association, London (Special Volume No. 15), xii + 633 pp.

—— and PANCHEN, A. L. 1973. Geographic variation in the tetrapod faunas of the Upper Carboniferous and Lower Permian. 353–368. *In* TARLING, D. H. and RUNCORN, S. K. (eds). *Implications of continental drift to the earth sciences*.

*Volume 1*. Academic Press, London and New York, 622 pp.

MORGAN, J. 1959. The morphology and anatomy of American species of the genus *Psaronius*. *Illinois Biological Monographs*, **27**, 1–108.

NEVES, R. and DOWNIE, C. (eds). 1967. *Geological excursions in the Sheffield region and the Peak District National Park*. University of Sheffield, 163 pp.

OWENS, B., RILEY, N. J. and CALVER, M. A. 1985. Boundary stratotypes and new stage names for the lower and middle Westphalian sequences in Britain. *Compte Rendu—10ᵉ Congrès de Stratigraphie et de Géologie du Carbonifère (Madrid, 1983)*, **4**, 461–472.

PANCHEN, A. L. 1970. *Handbuch der Paläoherpetologie. Teil 5a Anthracosauria*. Fischer, Stuttgart, 84 pp.

—— 1973. Carboniferous tetrapods. 117–125. *In* HALLAM, A. (ed.). *Atlas of palaeogeography*. Elsevier, Amsterdam, xii + 532 pp.

PATTEISKY, K. 1957. Die phylogenetische Entwicklung der Arten von *Lyginopteris* und ihre Bedeutung für die Stratigraphie. *Mitteilungen der Westfälischen Bergwerkschaftsklasse*, **12**, 59–83.

PEPPERS, R. A. and PFEFFERKORN, H. W. 1970. A comparison of the floras of the Colchester (No.2) Coal and Francis Creek Shale. *Illinois Geological Survey Guidebook Series*, **8**, 61–74.

PETRUNKEVITCH, A. 1953. Paleozoic and Mesozoic Arachnida of Europe. *Geological Society of America Memoir*, **53**, 1–128.

PHILLIPS, T. L. 1980. Stratigraphic and geographic occurrences of permineralized coal-swamp plants—Upper Carboniferous of North America. 25–92. *In* DILCHER, D. L. and TAYLOR, T. N. (eds). *Biostratigraphy of fossil plants. Successional and paleoecological analyses*. Dowden, Hutchinson and Ross Inc., Stroudsburg, 259 pp.

—— and DIMICHELE, W. A. 1992. Comparative ecology and life-history biology of arborescent lycopsids in Late Carboniferous swamps of Euramerica. *Annals of the Missouri Botanical Garden*, **79**, 560–588.

—— and PEPPERS, R. A. 1984. Changing patterns of Pennsylvanian coal-swamp vegetation and implications of climatic control on coal occurrence. *International Journal of Coal Geology*, **3**, 205–255.

—— —— and DIMICHELE, W. A. 1985. Stratigraphic and interregional changes in Pennsylvanian coal-swamp vegetation: environmental inferences. *International Journal of Coal Geology*, **5**, 43–109.

RAMSBOTTOM, W. H. C. 1979. Rates of transgression and regression in the Carboniferous of N.W. Europe. *Journal of the Geological Society, London*, **136**, 347–353.

—— CALVER, M. A., EAGAR, R. M. C., HODSON, F., HOLLIDAY, D. W., STUBBLEFIELD, C. J. and WILSON, R. B. 1978. *A correlation of Silesian rocks in the British Isles*. Geological Society, London (Special Report No. 10), 82 pp.

REX, G. 1983. The compression state of preservation of Carboniferous lepidodendrid leaves. *Review of Palaeobotany and Palynology*, **39**, 65–85.

—— and CHALONER, W. G. 1983. The experimental formation of plant compression fossils. *Palaeontology*, **26**, 231–252.

ROLFE, W. D. I. 1980. Early terrestrial faunas. 117–157. *In* PANCHEN, A. L. (ed.). *The terrestrial environment and the origin of land vertebrates*. Systematics Association, London (Special Volume No. 15), xii + 633 pp.

—— 1986. Aspects of the Carboniferous terrestrial arthropod community. *Compte Rendu—9ᵉ Congrès de Stratigraphie et de Géologie du Carbonifère (Washington and Urbana, 1979)*, **5**, 303–316.

ROTHWELL, G. W. 1975. The Callistophytaceae (Pteridospermopsida): I. Vegetative structures. *Palaeontographica, Abteilung B*, **151**, 171–196.

—— 1981. The Callistophytales (Pteridospermopsida). Reproductively sophisticated gymnosperms. *Review of Palaeobotany and Palynology*, **32**, 103–121.

—— 1988. Cordaitales. 273–297. *In* BECK, C. B. (ed.). *Origin and evolution of gymnosperms.* Columbia University Press, New York, 504 pp.

ROWLEY, D. B., RAYMOND, A., PARRISH, J. T., LOTTES, A. L., SCOTESE, C. R. and ZIEGLER, A. M. 1985. Carboniferous paleogeographic, phytogeographic, and paleoclimatic reconstructions. *International Journal of Coal Geology*, **5**, 7–42.

SAVAGE, R. J. G. (ed.). 1977. *Geological excursions in the Bristol district.* University of Bristol, 196 pp.

SCHEIHING, M. H. and PFEFFERKORN, H. W. 1984. The taphonomy of land plants in the Orinoco Delta: a model for the incorporation of plant parts in clastic sediments of Late Carboniferous age of Euramerica. *Review of Palaeobotany and Palynology*, **41**, 205–240.

SCHNEIDER, J. 1983. Die Blattoidea (Insecta) des Palaeozoikums. Teil 1: Systematik, Oekologie und Biostratigraphie. *Freiberger Forschungshefte*, **C382**, 106–146.

SCOTT, A. C. 1977. A review of the ecology of Upper Carboniferous plant assemblages, with new data from Strathclyde. *Palaeontology*, **20**, 447–473.

—— 1978. Sedimentological and ecological control of Westphalian B plant assemblages from West Yorkshire. *Proceedings of the Yorkshire Geological Society*, **41**, 451–508.

—— 1979. The ecology of Coal Measure floras from northern Britain. *Proceedings of the Geologists' Association*, **90**, 97–116.

—— 1980. The ecology of some Upper Palaeozoic floras. 87–115. *In* PANCHEN, A. L. (ed.). *The terrestrial environment and the origin of land vertebrates.* Systematics Association, London (Special Volume No. 15), xii + 633 pp.

—— 1984. Studies on the sedimentology, palaeontology and palaeoecology of the Middle Coal Measures (Westphalian B, Upper Carboniferous) at Swillington, Yorkshire. Part 1. *Transactions of the Leeds Geological Association*, **10**, 1–16.

—— 1989. Observations on the nature and origin of fusain. *International Journal of Coal Geology*, **12**, 443–475.

—— and CHALONER, W. G. 1983. The earliest fossil conifer from the Westphalian B of Yorkshire. *Proceedings of the Royal Society of London, Series B*, **220**, 163–182.

—— and COLLINSON, M. E. 1978. Organic sedimentary particles: results from scanning electron microscope studies of fragmentary plant material. 137–167. *In* WHALLEY, W. B. (ed.). *Scanning electron microscopy in the study of sediments.* Geo Abstracts, Norwich, v + 414 pp.

—— and REX, G. 1985. The formation and significance of Carboniferous coal balls. *Philosophical Transactions of the Royal Society of London, Series B*, **311**, 123–137.

—— and TAYLOR, T. N. 1983. Plant/animal interactions during the Upper Carboniferous. *Botanical Review*, **49**, 259–307.

SHUTE, C. H. and CLEAL, C. J. 1987. Palaeobotany in museums. *Geological Curator*, **4**, 553–559.
—— —— 1989. The holotype of the Carboniferous marattialean fern *Lobatopteris miltoni* (Artis). *Bulletin of the British Museum (Natural History), Geology Series*, **45**, 71–76.
SMART, J. and HUGHES, N. F. 1973. The insect and the plant: progressive palaeoecological integration. 143–155. *In* VAN EMDEN, H. F. (ed.). *Insect/plant relationships. Royal Entomological Society, Symposium No. 6.* Blackwell, London, vii + 215 pp.
SPICER, R. A. 1989. The formation and interpretation of plant fossil assemblages. *Advances in Botanical Research*, **16**, 95–191.
STEWART, W. N. and DELEVORYAS, T. 1956. The medullosan pteridosperms. *Botanical Review*, **22**, 45–80.
STIDD, B. M. 1981. The current status of medullosan seed ferns. *Review of Palaeobotany and Palynology*, **32**, 63–101.
STOCKMANS, F. and WILLIÈRE, Y. 1961. *Végétaux du Westphalien A de la Belgique. Graines, inflorescences et synanges.* Centre National de Géologie Houillère, Brussels (Publication No. 4), 118 pp.
STORCH, D. 1966. Die Arten der Gattung *Sphenophyllum* Brongniart im Zwickau-Lugau-Oelsnitzer Steinkohlenrevier. *Paläontologische Abhandlungen, B*, **2**, 193–426.
STUBBLEFIELD, C. J. and TROTTER, F. M. 1957. Divisions of the Coal Measures on Geological Survey maps of England and Wales. *Bulletin of the Geological Survey of Great Britain*, **13**, 1–5.
TAYLOR, T. N. 1981. *Paleobotany. An introduction to fossil plant biology.* McGraw-Hill, New York, 589 pp.
—— and MILLAY, M. A. 1981. Morphologic variability of Pennsylvanian lyginopterid seed ferns. *Review of Palaeobotany and Palynology*, **32**, 27–62.
—— and SCOTT, A. C. 1983. Interactions of plants and animals during the Carboniferous. *BioScience*, **33**, 488–493.
THOMAS, B. A. 1966. The cuticle of the Lepidodendroid stem. *New Phytologist*, **65**, 296–303.
—— 1967a. *Ulodendron* Lindley and Hutton and its cuticle. *Annals of Botany*, **31**, 775–782.
—— 1967b. The cuticle of two species of *Bothrodendron* (Lycopsida; Lepidodendrales). *Journal of Natural History*, **1**, 53–60.
—— 1968. The Carboniferous fossil lycopod *Ulodendron landsburgii* (Kidston) comb. nov. *Journal of Natural History*, **2**, 425–428.
—— 1970a. Epidermal studies in the interpretation of *Lepidodendron* species. *Palaeontology*, **13**, 145–173.
—— 1970b. A new specimen of *Lepidostrobus binneyanus* from the Westphalian B of Yorkshire. *Pollen et Spores*, **12**, 217–234.
—— 1972. Growth changes in *Sigillaria latibasa*. *Annals of Botany*, **36**, 1023–1027.
—— 1977. Epidermal studies in the interpretation of *Lepidophloios* species. *Palaeontology*, **20**, 273–293.
—— 1978. Carboniferous Lepidodendraceae and Lepidocarpaceae. *Botanical Review*, **44**, 321–364.

—— 1981. Structural adaptations shown by the Lepidocarpaceae. *Review of Palaeobotany and Palynology*, **32**, 377–388.

—— 1987a. The formation of large diameter plant fossil moulds and the Walton theory of compaction. *Geological Journal*, **21**, 381–385.

—— 1987b. The use of *in-situ* spores for defining species of dispersed spores. *Review of Palaeobotany and Palynology*, **51**, 227–233.

—— 1988. The fine structure of the Carboniferous lycophyte microspore *Lycospora perforata* Bharadwaj and Venkatachala. *Pollen et Spores*, **30**, 81–88.

—— and BLACKBURN V. 1987. An ultrastructural study of some Carboniferous *in situ* megaspores assignable to *Lagenicula horrida* and *Lagenoisporites rugosus*. *Pollen et Spores*, **29**, 435–448.

—— and CLEAL, C. J. 1993. Middle Westphalian plant fossils from the West Cumberland Coalfield, Great Britain. *Geological Journal*, **28**, 101–123.

—— —— 1993a. *The Coal Measure Forests*. National Museum of Wales, Cardiff, 32 pp.

—— —— 1994. Plant fossils from the Writhlington Geological Nature Reserve. *Proceedings of the Geologists' Association*, **105**, 15–32.

—— and DYTKO A. 1980. *Lepidostrobus haslingdenensis*: a new species from the Lancashire Millstone Grit. *Geological Journal*, **15**, 137–142.

—— and MASARATI, D. L. 1982. Cuticular and epidermal studies in fossil and living lycophytes. 363–387. *In* CUTLER, D. F., ALVIN, K. L. and PRICE, C. E. (eds). *The plant cuticle*. Linnean Society, London (Symposium Series No. 10), 461 pp.

—— and SPICER, R. A. 1987. *The evolution and palaeobiology of land plants*. Croom Helm, London and Sydney, 318 pp.

—— and WATSON, J. 1976. A rediscovered 114–foot *Lepidodendron* from Bolton, Lancashire. *Geological Journal*, **11**, 15–20.

TRIVETT, M. L. and ROTHWELL, G. W. 1991. Diversity among Paleozoic Cordaitales. *Neues Jahrbuch für Geologie und Paläontologie, Abhandlungen*, **183**, 289–305.

VAKHRAMEEV, V. A. 1991. *Jurassic and Cretaceous floras and climates of the Earth*. Cambridge University Press, Cambridge, 318 pp.

WAGNER, R. H. 1961. Some Alethopterideae from the South Limburg Coalfield. *Mededelingen van de Geologische Stichting, Nieuwe Serie*, **14**, 5–13.

—— 1965. Stephanian B flora from the Ciñera-Matallana coalfield (León) and neighbouring outliers. III. *Callipteridium* and *Alethopteris*. *Notas y Comunicaciones del Instituto Geológico y Minero de España*, **78**, 5–70.

—— 1968. Upper Westphalian and Stephanian species of *Alethopteris* from Europe, Asia Minor and North America. *Mededelingen Rijks Geologische Dienst, Serie C*, **3–1–6**, 1–188.

—— 1974. The chronostratigraphic units of the Upper Carboniferous in Europe. *Bulletin Société Belge de Géologie*, **83**, 235–253.

—— 1984. Megafloral zones of the Carboniferous. *Compte Rendu—9ᵉ Congrès de Stratigraphie et de Géologie du Carbonifère (Washington and Urbana, 1979)*, **2**, 109–134.

—— and ALVAREZ-VÁZQUEZ, C. 1991. Floral characterisation and biozones of the Westphalian D Stage in NW Spain. *Neues Jahrbuch für Geologie und Paläontologie, Abhandlungen*, **183**, 171–202.

—— FERNANDEZ GARCIA, L. G. and EAGAR, R. M. C. 1983. Geology and *palaeontology of the Guardo Coalfield (NE León—NW Palencia).* Instituto Geológico y Minero de España, Madrid, 109 pp.

—— and SPINNER, E. 1972. The stratigraphic implications of the Westphalian D macro- and microflora of the Forest of Dean Coalfield (Gloucestershire), England. *Proceedings of the International Geological Congress (Montreal, 1972),* **7**, 428–431.

—— and WINKLER PRINS, C. F. 1985. Stratotypes of the two lower Stephanian stages, Cantabrian and Barruelian. *Compte Rendu — 10$^e$ Congrès de Stratigraphie et de Géologie du Carbonifère (Madrid, 1983),* **4**, 473–483.

WALTON, J. 1933. The palæobotany of Robin Hood Quarry, Leeds. *Naturalist, Hull,* **1933**, 201–205.

WEISS, E. 1876. Steinkohlen-Calamarien, I. *Abhandlungen zur geologische Specialkarte von Preussen und den Thüringischen Staaten,* **2**(1), 1–149.

WINKLER PRINS C. F. 1989. Summary chart of biostratigraphic levels proposed for intercontinental correlation of the Upper Carboniferous (Pennsylvanian). *Newsletter on Carboniferous Stratigraphy,* **7**, 16.

WOODLAND, A. W. and EVANS, J. V. 1964. The country around Pontypridd and Maesteg. *Memoirs of the Geological Survey of Great Britain,* 391 pp.

ZODROW, E. L. 1985. *Odontopteris* Brongniart in the Upper Carboniferous of Canada. *Palaeontographica, Abteilung B,* **196**, 79–110.

—— and CLEAL, C. J. 1985. Phyto- and chronostratigraphical correlations between the late Pennsylvanian Morien Group (Sydney, Nova Scotia) and the Silesian Pennant Measures (south Wales). *Canadian Journal of Earth Sciences,* **22**, 1465–1473.

—— —— 1988. The structure of the Carboniferous pteridosperm frond *Neuropteris ovata* Hoffmann. *Palaeontographica, Abteilung B,* **208**, 105–124.

—— —— 1993. The epidermal structure of the Carboniferous gymnosperm frond *Reticulopteris. Palaeontology,* **36**, 65–79.

# TAXONOMIC INDEX

This index lists all of the form-species and form-genera referred to in the key. Each form-genus is given in alphabetical order, with its author citation and a note as to the type of plant organ that each belongs to. The species are then listed for each form-genus, again with author citation, which was omitted from the body of the key for brevity. It also refers to the key factor(s) for each species. For instance, G-19 indicates that it keys out in Group G at couplet 19. Finally, reference is given to an illustration in this work, if given.

Species

Key/Figure

*Alethopteris* Sternberg, 1825 (Pteridosperm leaves)
*A. bertrandii* Buisine, 1961 ...................................E-45 (Text-fig. 59A)
*A. davreuxii* (Brongniart) Zeiller, 1886 ..................E-47 (Text-fig. 61D)
*A. decurrens* (Artis) Zeiller, 1886 .....E-47 (Pl. 21, fig. 3; Text-fig. 61C)
*A. grandinioides* Kessler, 1916 ..................................E-51, E-52
.............................................................(Pl. 19, fig. 2; Text-fig. 60A,B)
*A. lancifolia* Wagner, 1961 ....................................E-53 (Text-fig. 59C)
*A. lesquereuxii* Wagner, 1968 ...........E-52 (Pl. 21, fig. 2; Text-fig. 60D)
*A. lonchitica* Sternberg, 1826 ................................E-54 (Text-fig. 60E)
*A. robusta* Lesquereux, 1884 ................................E-45 (Text-fig. 60C)
*A. serlii* (Brongniart) Göppert, 1836 ........................E-47, E-51
.............................................................(Pl. 21, fig. 1; Text-fig. 59B)
*A. urophylla* (Brongniart) Presl, 1838 .........E-54 (Pl. 20; Text-fig. 61A)
*A. valida* Boulay, 1876 .........................................E-50 (Text-fig. 61B)

*Annularia* Sternberg, 1821 (Calamite leaves)
*A. galioides* (Lindley and Hutton) Kidston, 1891......C-9 (Text-fig. 44D)
*A. mucronata* Schenk, 1883 ....................................C-11 (Text-fig. 44B)
*A. radiata* (Brongniart) Sternberg, 1825 ...................C-9 (Text-fig. 44C)
*A. sphenophylloides* (Zenker) Gutbier, 1837 ...........C-11 (Text-fig. 44E)
*A. stellata* (Sternberg) Wood, 1861 ..............C-10 (Pl. 14; Text-fig. 44A)

*Aphlebia* Presl, 1838 (Part of fern leaf)
*A. crispa* (Gutbier) Presl, 1838 ..............................E-57 (Text-fig. 63)

*Artisia* Sternberg, 1838 (Cordaite stem pith cast) .........A-2 (Text-fig. 40B)

*Artisophyton* Pfefferkorn, 1976 (Arborescent fern stems)
*A. goldenbergii* (Weiss) Pfefferkorn, 1976..............B-68 (Text-fig. 42A)

*A. approximatum* (Lindley and Hutton) Pfefferkorn, 1976............B-68
..................................................................................(Text-fig. 42B)

*Asolanus* Wood, 1861 (Arborescent lycopsid stems)
 *A. camptotaenia* Wood, 1861..........................................B-4 (Text-fig. 28B)

*Asterophyllites* Brongniart, 1822 (Calamite leaves)
 *A. charaeformis* (Sternberg) Unger, 1845..................C-7 (Text-fig. 43C)
 *A. equisetiformis* Brongniart, 1828......C-5 (Pl. 13, fig. 2; Text-fig. 43A)
 *A. grandis* (Sternberg) Geinitz, 1855......................C-6 (Text-fig. 43E)
 *A. longifolius* (Sternberg) Brongniart, 1828.................................C-4
 ....................................................................(Pl. 13, fig. 3; Text-fig. 43D)
 *A. lycopodioides* Zeiller, 1886....................................C-7 (Text-fig. 43B)

*Aulacotheca* Halle, 1933 (Pteridosperm pollen organs)
 *A. elongata* (Kidston) Halle, 1933.......................H-39 (Text-fig. 91E)
 *A. hallei* Hemingway, 1941...............H-39 (Pl. 29, fig. 8; Text-fig. 91F)
 *A. hemingwayi* Halle, 1933....................................H-38 (Text-fig. 91D)

*Bertrandia* Dalinval, 1960 (Fern fronds)
 *B. avoldensis* (Stur) Dalinval, 1960......................E-124 (Text-fig. 77A)

*Bothrodendron* Lindley and Hutton, 1833 (Arborescent lycopsid stems)
 *B. minutifolium* Boulay, 1876....................................................B-58, D-13
 ........................................................(Pl. 6, fig. 4; Text-figs 38A-C, 47C,D)
 *B. punctatum* Lindley and Hutton, 1833.......B-58, D-13 (Text-fig. 38D)

*Boulaya* Halle, 1933 (Isolated pteridosperm organs)
 *B. fertilis* (Kidston) Halle, 1933.........H-37 (Pl. 29, fig. 2; Text-fig. 91C)
 *B. praelonga* Carpentier, 1934...............................H-37 (Text-fig. 91B)

*Bowmanites* Binney, 1871 (Sphenophyte cones)
 *B.* sp. ..........................................................................................F-5

*Calamites* Suckow ex Brongniart, 1828
(Sphenopsid stems and pith casts)
 *C. brongniartii* Sternberg, 1833...............................A-8 (Text-fig. 25D)
 *C. carinatus* Sternberg, 1823....................................A-7 (Text-fig. 24)
 *C. cistii* Brongniart, 1828........................................A-5 (Text-fig. 23A)
 *C. goeppertii* Ettingshausen, 1854.........A-9 (Pl. 2, fig. 2; Text-fig. 25A)
 *C. multiramis* Weiss, 1884....................A-8 (Pl. 2, fig. 1; Text-fig. 23C)
 *C. schuetzeiformis* Kidston and Jongmans, 1917 ....A-10 (Text-fig. 25B)
 *C. suckowii* Brongniart, 1828.................A-5 (Pl. 1, fig. 2; Text-fig. 23B)
 *C. undulatus* Sternberg, 1825...............................A-10 (Text-fig. 25C)

Taxonomic Index 213

*Calamostachys* Schimper, 1869 (Sphenopsid cones)
 *C. germanica* Weiss, 1876 ................................F-8 (Text-fig. 83D)
 *C. northumbriana* (Kidston) Jongmans, 1911 ...........F-9 (Text-fig. 83E)
 *C. paniculata* Weiss, 1876 ................................F-11 (Text-fig. 83C)
 *C. ramosa* Weiss, 1884 ....................F-10 (Pl. 15, fig. 5; Text-fig. 83B)
 *C. striata* (Weiss) Hirmer, 1927 .........................F-8 (Text-fig. 83A)
 *C. tuberculata* (Sternberg) Weiss, 1884 ................F-11 (Text-fig. 83F)

*Callipteridium* (Weiss) Zeiller, 1888 (Pteridosperm leaves)
 *C. armasii* (Zeiller) Wagner, 1963 ......................E-42 (Text-fig. 58D)
 *C. jongmansii* (Bertrand) Wagner, 1958 ................E-42 (Text-fig. 58C)

*Carpolithus* Linnaeus ex Sternberg, 1825 (Isolated seeds)
 *C. areolatus* (Boulay) Boulay, 1911 ....................H-31 (Text-fig. 90B)
 *C. granularis* Sternberg, 1826 ..........................H-32 (Text-fig. 90A)
 *C. inflatus* (Lesquereux) Kidston, 1890 ................H-28 (Text-fig. 90G)
 *C. membranaceous* Göppert, 1864 ............H-31, H-33 (Text-fig. 90C)
 *C. perpusillus* Lesquereux, 1884 ........................H-33 (Text-fig. 90E)
 *C. pseudosulcatus* Crookall, 1976 .......................H-30 (Text-fig. 90D)
 *C. wildii* Kidston, 1891 ...............H-28 (Pl. 29, fig. 10; Text-fig. 90F)

*Caulopteris* Lindley and Hutton, 1833 (Arborescent fern stems)
 *C. anglica* Kidston, 1888 ....................B-64 (Pl. 11; Text-fig. 41A)
 *C. arberi* Crookall, 1955 ................................B-65 (Text-fig. 41C)
 *C. cyclostigma* (Lesquereux) Kidston, 1901 ...........B-65 (Text-fig. 41B)

*Cordaicarpus* Geinitz, 1862 (Isolated cordaite seeds)
 *C. congruens* Grand'Eury, 1877 .........................H-12 (Text-fig. 87D)
 *C. cordai* (Geinitz) Geinitz, 1862 ......H-12 (Pl. 29, fig. 7; Text-fig. 87C)

*Cordaitanthus* Feistmantel, 1876 (Cones and isolated seeds of cordaites)
 *C. lindleyi* (Carruthers) Renault, 1881 .................H-9 (Text-fig. 87G)
 *C. pseudofluitans* Kidston, 1922 ........................H-9 (Text-fig. 87F)
 *C.* sp. ..........................................................F-1 (Text-fig. 80)

*Cordaites* Unger, 1850 (Cordaite leaves)
 *C. angulosostriatus* Grand'Eury, 1877 ..................A-17 (Text-fig. 27B)
 *C. borassifolius* (Sternberg) Unger, 1850 .............................A-16
 *C. microstachys* Goldenberg, 1869 .........................A-14, A-16
 *C. palmaeformis* (Göppert) Weiss, 1871 ..............................A-14
 *C. principalis* (Germar) Geinitz, 1855.A-17 (Pl. 1, fig. 1; Text-fig. 27A)

*Cornucarpus* Crookall, 1976 (Isolated seeds)
 *C. acutus* (Lindley and Hutton) Crookall, 1976 .......H-6 (Text-fig. 87B)
 *C. arberi* Crookall, 1976 ................................H-6 (Text-fig. 87A)

*Corynepteris* Baily, 1860 (Fern fronds)
 *C. angustissima* (Sternberg) Němejc, 1938 ............................................. E-95
 .................................................. (Pl. 27, fig. 1; Text-fig. 73C; Pl. 17, fig. 2)
 *C. coralloides* (Gutbier) Bertrand, 1914 .................. E-95 (Text-fig. 73B)
 *C. similis* (Sternberg) Kidston, 1911 ........................ E-94 (Text-fig. 73A)

*Crossotheca* Zeiller, 1833 (Fern fronds)
 *C. crepinii* Zeiller, 1883 ............................. E-103, E-106 (Text-fig. 74G)

*Cyathocarpus* Weiss, 1869 (Fern fronds)
 *C. hemitelioides* (Brongniart) Mosbrugger, 1983...E-132 (Text-fig. 79F)
 *C.* aff. *arborescens* (Brongniart) Weiss, 1869 ..................... E-131, E-132
 ................................................................... (Pl. 27, fig. 2; Text-fig. 79G)

*Cyclopteris* Brongniart, 1828 (Basal parts of pteridosperm leaves)
 *C. fimbriata* Lesquereux, 1858 ................................. E-9 (Text-fig. 51B)
 *C. orbicularis* Brongniart, 1828 ............ E-9 (Pl. 17, fig. 1; Text-fig. 51A)

*Cyclostigma* Haughton, 1860 (Lycopsid stems)
 *C. cambricum* Crookall, 1964 ................................. B-59 (Text-fig. 39A)
 *C. macconochiei* Kidston, 1903 ............................. B-59 (Text-fig. 39B)

*Cyperites* Lindley and Hutton, 1832 (Lycopsid leafy shoots)
 *C. bicarinatus* Lindley and Hutton, 1832 ............................. A-12, G-2
 .......................................................... (Pl. 5, fig. 3; Text-fig. 26B)
 *C. ciliatus* Crookall, 1966 ........................... A-12, G-2 (Text-fig. 26A)

*Desmopteris* Stur, 1883 (?Pteridosperm leaves)
 *D. longifolia* (Presl) Potonié, 1904 ......................... E-40 (Text-fig. 58B)

*Dicksonites* Sterzel, 1881 (Pteridosperm leaves)
 *D. plueckenetii* (Sternberg) Sterzel, 1881 .................. E-83 (Text-fig. 72C)

*Eusphenopteris* (Gothan) Simson-Scharold, 1934 (Pteridosperm leaves)
 *E. foliolata* (Stur) van Amerom, 1975 ..................... E-69 (Text-fig. 68A)
 *E. grandis* (Keller) van Amerom, 1975 .................. E-74 (Text-fig. 69A)
 *E. hollandica* (Gothan and Jongmans) Novik, 1947 ..............................
 ............................................................................. E-75 (Text-fig. 69C)
 *E. neuropteroides* (Boulay) Novik, 1947 ................................... E-70
 ............................................................... (Pl. 26, fig. 2; Text-fig. 67)
 *E. nummularia* (Gutbier) Novik, 1947 ........................................ E-76
 ................................................................ (Pl. 25, fig. 1; Text-fig. 69D)
 *E. obtusiloba* (Brongniart) Novik, 1947 ................. E-72 (Text-fig. 66A)
 *E. sauveurii* (Crépin) Simson-Scharold, 1934 ............................. E-68
 ............................................................... (Pl. 26, fig. 3; Text-fig. 66D)

Taxonomic Index 215

*E. scribanii* van Amerom, 1975 ............................E-74 (Text-fig. 68B)
*E. striata* (Gothan) Novik, 1947 ...........................E-76 (Text-fig. 66C)
*E. trifoliolata* (Artis) Novik, 1947 .....E-71 (Pl. 26, fig. 1; Text-fig. 69B)
*E. trigonophylla* (Behrend) van Amerom, 1975 ......E-72 (Text-fig. 66B)

*Flemingites* (Carruthers) Brack-Hanes and Thomas, 1983 (Lycopsid cone)
*F.* sp. .................................................................................F-3

*Fortopteris* Boersma, 1969 (Pteridosperm leaves)
*F. latifolia* (Brongniart) Boersma, 1969 ..................E-65 (Text-fig. 71B)

*Gnetopsis* Kidston *in* Seward, 1917 (Isolated pteridosperm seeds)
*G. anglica* Kidston *in* Seward, 1917 .....................H-16 (Text-fig. 88A)

*Hexagonocarpus* Renault, 1896 (Isolated pteridosperm seeds)
*H. hookeri* Kidston, 1914 .......................................H-22 (Text-fig. 89G)

*Holcospermum* Nathorst, 1914 (Isolated pteridosperm seeds)
*H. mammillatum* (Lesquereux) Crookall, 1976 ......H-20 (Text-fig. 88C)
*H. mentzelianum* (Göppert) Crookall, 1976...........H-20 (Text-fig. 88B)
*H. multistriatum* (Presl) Crookall, 1976 .................H-21 (Text-fig. 88D)
*H. sulcatum* (Presl) Seward, 1917 .....H-21 (Pl. 29, fig. 9; Text-fig. 88E)

*Hymenophyllites* Zeiller, 1883 (Fern fronds)
*H. quadridactylites* (Gutbier) Kidston, 1923..........E-100 (Text-fig. 73E)

*Karinopteris* Boersma, 1972 (Pteridosperm leaves)
*K. acuta* (Brongniart) Boersma, 1972...................................E-81
................................................................. (Pl. 24, fig. 1; Text-fig. 70C)
*K. daviesii* (Kidston) Boersma, 1972 .....................E-79 (Text-fig. 70B)
*K. grandepinnata* (Huth) Boersma, 1972 ...............E-78 (Text-fig. 70E)
*K. nobilis* (Achepohl) Boersma, 1972....................E-81 (Text-fig. 70D)
*K. robusta* (Danzé-Corsin) Boersma, 1972............E-78 (Text-fig. 71C)
*K. soubeiranii* (Zeiller) Boersma, 1972.................E-80 (Text-fig. 70A)

*Lagenospermum* (Isolated pteridosperm seeds)
*L.* sp. ....................................................................H-16 (Text-fig. 87E)

*Laveineopteris* Cleal *et al.*, 1990 (Pteridosperm leaves)
*L. loshii* (Brongniart) Cleal *et al.*, 1990 ...............E-23 (Text-fig. 53C,D)
*L. rarinervis* (Bunbury) Cleal *et al.*, 1990...........................E-20
..................................................................(Pl. 17, fig. 4; Text-fig. 53B)
*L. tenuifolia* (Sternberg) Cleal *et al.*, 1990..........................E-25
..................................................................(Pl. 17, fig. 5; Text-fig. 53A)

# British Coal Measures

*Lepidodendron* Sternberg, 1820
(Arborescent lycopsid stems and leafy shoots)
*L. aculeatum* Sternberg, 1820 ............... B-18, B-22 (Pl. 3; Text-fig. 33)
*L. acutum* (Presl) Kidston, 1911 ..................... D-11 (Text-figs 48A, 49)
*L. arberi* Thomas, 1970 ........................................... B-19 (Text-fig. 30A)
*L. barnsleyense* Thomas, 1970 ................................. B-11 (Text-fig. 29E)
*L. dichotomum* Sternberg, 1820 ........... B-10 (Pl. 4, fig. 2; Text-fig. 29A)
*L. feistmantelii* Zalessky, 1875 ......... B-14 (Pl. 4, fig. 5; Text-fig. 30C,D)
*L. fusiforme* (Corda) Unger, 1850 .......................... B-19 (Text-fig. 31A)
*L. jaraczewskii* Zeiller, 1886 ............... B-14 (Pl. 4, fig. 1; Text-fig. 31B)
*L. lycopodioides* Sternberg, 1821 ..................................................... D-12
.................................................. (Pl. 4, fig. 4; Pl. 15, fig. 4; Text-fig. 48E)
*L. mannabachense* Presl, 1838 ........................................................ B-11
................................................... (Pl. 4, fig. 6; Pl. 5, fig. 2; Text-fig. 29B,C)
*L. ophiurus* Brongniart, 1822 ................ D-9 (Pl. 6, fig. 5; Text-fig. 48C)
*L. peachii* Kidston, 1885 ............................................. B-8 (Text-fig. 29D)
*L. rimosum* Sternberg, 1820 ................ B-20 (Pl. 5, fig. 1; Text-fig. 31C)
*L. serpentigerum* Koenig, 1825 ............................. B-22 (Text-fig. 32D)
*L. simile* Kidston, 1909 .......................................... D-11 (Text-fig. 48B)
*L. subdichotum* Sterzel, 1901 ................................. B-12 (Text-fig. 30B)
*L. volkmanianum* Sternberg, 1825 ........................ B-17 (Text-fig. 32B)
*L. wedekindii* Weiss, 1893 ..................................... B-21 (Text-fig. 32A)
*L. worthenii* Lesquereux, 1866 ................................................ B-17, D-9
.................................................................... (Pl. 6, fig. 1; Text-figs 32C, 48D)

*Lepidophloios* Sternberg, 1825 (Arborescent lycopsid stems)
*L. acerosus* Lindley and Hutton, 1831 ................................. B-15, B-24
.............................................................. (Pl. 4, fig. 3; Text-fig. 34A,B)
*L. laricinus* Sternberg, 1825 ................. B-25 (Pl. 6, fig. 2; Text-fig. 34C)
*L. macrolepidotus* Goldenberg, 1862 ... B-25 (Pl. 6, fig. 3; Text-fig. 34D)

*Lepidostrobophyllum* (Hirmer) Allen, 1966 (Isolated lycopsid sporophylls)
*L. acuminatum* (Lesquereux) Bell, 1940 ................... G-9 (Text-fig. 84H)
*L. alatum* Boulter, 1968 ................. G-6, G-7 (Text-figs 84D-F,J,K, 85)
*L. hastatum* (Lesquereux) Chaloner, 1967 ............. G-4 (Text-fig. 84A,C)
*L. lanceolatum* (Lindley and Hutton) Bell, 1938 ...... G-9 (Text-fig. 84G)
*L. majus* (Brongniart) Hirmer, 1927 ......................... G-7 (Text-fig. 84I)
*L. triangulare* (Zeiller) Bell, 1938 ........................... G-8 (Text-fig. 84B)
*L.* sp. ............................................................................................... C-2

*Lepidostrobus* Brongniart, 1828 (Lycopsid cone)
*L.* sp. .................................................................... F-3 (Pl. 28, fig. 1)

*Linopteris* Presl, 1838 (Pteridosperm leaves)
*L. bunburii* Bell, 1962 ................................................E-36 (Text-fig. 57C)
*L. neuropteroides* (Gutbier) Zeiller, 1899 ...............E-37 (Text-fig. 57A)
*L. subbrongniartii* (Grand'Eury) Carpentier, 1911..E-37 (Text-fig. 57B)

*Lobatopteris* Wagner, 1958 (Fern fronds)
*L. micromiltoni* (Corsin) Wagner, 1958 ................E-122 (Text-fig. 77C)
*L. miltoni* (Artis) Wagner, 1958 .............................E-124 (Text-fig. 78B)
*L. vestita* (Lesquereux) Wagner, 1958 ...................E-121 (Text-fig. 77D)

*Lonchopteris* Brongniart, 1828 (Pteridosperm leaves)
*L. eschweileriana* Andrä, 1865 ................................E-56 (Text-fig. 62A)
*L. rugosa* Brongniart, 1835 ................E-56 (Pl. 19, fig. 4; Text-fig. 62C)
*L. petitii* Buisine, 1961 ............................................E-55 (Text-fig. 62B)

*Lycopodites* Lindley and Hutton, 1833 (Herbaceous lycopsid shoots)
*L. pendulus* Lesquereux, 1879 ................................................................D-5

*Lyginopteris* Potonié, 1897 (Pteridosperm leaves)
*L. baeumleri* (Andrä) Gothan, 1931 ........................E-92 (Text-fig. 72B)
*L. hoeninghausii* (Brongniart) Gothan, 1931 ..........E-91 (Text-fig. 72A)

*Macroneuropteris* Cleal *et al.*, 1990 (Pteridosperm leaves)
*M. macrophylla* (Brongniart) Cleal *et al.*, 1990 ......E-31 (Text-fig. 50B)
*M. scheuchzeri* (Hoffmann) Cleal *et al.*, 1990 .......E-8, E-10, E-30, E-39
............................................................(Pl. 18, figs 1,2; Text-fig. 50A)

*Macrostachya* Schimper, 1869 (Sphenopsid cones)
*M. infundibuliformis* (Brongniart) Schimper, 1869 ...F-4 (Text-fig. 81A)

*Margaritopteris* Gothan, 1913 (Pteridosperm leaves)
*M. conwayii* (Lindley and Hutton) Crookall, 1929..E-92 (Text-fig. 72D)

*Mariopteris* Zeiller, 1879 (Pteridosperm leaves)
*M. hirsuta* Corsin, 1932 ...........................................E-87 (Text-fig. 71A)
*M. hirta* (Stur) Kidston, 1925 ..................................E-87 (Text-fig. 71F)
*M. lobatifolia* Kidston, 1925 ....................................E-85 (Text-fig. 71H)
*M. muricata* (Brongniart) Zeiller, 1879 ............................E-85, E-89
..................................................................(Pl. 23; Text-fig. 71E)
*M. nervosa* (Brongniart) Zeiller, 1879 ...............................................E-88
..........................................................(Pl. 22, fig. 2; Text-fig. 71G)
*M. sauveurii* (Brongniart) Frech, 1899...........E-15, E-89 (Text-fig. 71D)

*Megaphyton* Artis, 1825 (Arborescent fern stems)
  *M. frondosum* Artis, 1825..........................................B-67 (Text-fig. 41E)
  *M. gwynnevaughanii* Crookall, 1955......................B-67 (Text-fig. 41D)

*Neuralethopteris* Cremer ex Laveine, 1967 (Pteridosperm leaves)
  *N. jongmansii* Laveine, 1967....................................E-17 (Text-fig. 53F)
  *N. rectinervis* (Kidston) Laveine, 1967..................E-17 (Text-fig. 53G)
  *N. schlehanii* (Stur) Laveine, 1967..........................E-16 (Text-fig. 53E)

*Neuropteris* (Brongniart) Sternberg, 1825 (Pteridosperm leaves)
  *N. dussartii* Laveine, 1967................E-23 (Pl. 17, fig. 3; Text-fig. 54D)
  *N. flexuosa* Sternberg, 1826...............E-32 (Pl. 22, fig. 1; Text-fig. 54A)
  *N. heterophylla* (Brongniart) Sternberg, 1825.......................E-27
  ..............................................................................(Pl. 16; Text-fig. 55A)
  *N. hollandica* Stockmans, 1933...............................E-25 (Text-fig. 54E)
  *N. jongmansii* Crookall, 1959............E-24 (Pl. 18, fig. 4; Text-fig. 55B)
  *N. obliqua* (Brongniart) Zeiller, 1886......................E-28 (Text-fig. 54C)
  *N. ovata* Hoffmann, 1826...........................................E-32 (Text-fig. 54B)
  *N. semireticulata* Josten, 1962............E-29 (Pl. 19, fig. 1; Text-fig. 55C)

*Noeggerathia* Sternberg, 1822 (?Progymnosperm leaves)
  *N. chalardii* Danzé, 1957............................................E-12 (Text-fig. 52A)
  *N. foliosa* Sternberg, 1822.........................................E-12 (Text-fig. 52B)

*Odontopteris* (Brongniart) Sternberg, 1825 (Pteridosperm leaves)
  *O. cantabrica* Wagner, 1969..............E-39 (Pl. 19, fig. 3; Text-fig. 58A)

*Oligocarpia* Göppert, 1841 (Fern fronds)
  *O. brongniartii* Stur, 1883........................................E-106 (Text-fig. 74F)
  *O. gutbieri* Göppert, 1841........................................E-105 (Text-fig. 74E)

*Palaeostachya* Weiss, 1876 (Sphenopsid cones)
  *P. elongata* (Presl) Weiss, 1876................................F-15 (Text-fig. 82D)
  *P. ettingshausenii* Kidston, 1903..............................F-14 (Text-fig. 82E)
  *P. gracillima* Weiss, 1884...........................................F-15 (Text-fig. 82F)
  *P. minuta* Kidston, 1914.............................................F-13 (Text-fig. 82C)
  *P. paucibracteata* Sandberger, 1890........................F-13 (Text-fig. 82A)
  *P. pedunculata* Williamson ex Weiss, 1884............F-15 (Text-fig. 82B)

*Palmatopteris* Potonié, 1891 (Pteridosperm leaves)
  *P. furcata* (Brongniart) Potonié, 1891...................................E-62
  ....................................................... (Pl. 24, fig. 2; Text-fig. 64A)
  *P. geniculata* (Germar and Kaulfuss) Potonié, 1891.......................E-62
  ......................................................................(Pl. 25, fig. 3; Text-fig. 64B)
  *P. sturii* Gothan, 1913...............................................E-61 (Text-fig. 64C)

Taxonomic Index  219

*Paripteris* Gothan, 1941 (Pteridosperm leaves)
*P. gigantea* (Sternberg) Gothan, 1941 ............E-34 (Text-fig. 56B)
*P. linguaefolia* (Bertrand) Laveine, 1967 ............E-35 (Text-fig. 56A)
*P. pseudogigantea* (Potonié) Gothan, 1953 ............E-35
............(Pl. 18, fig. 3; Text-fig. 56C)
*P.* sp. ............E-8

*Pecopteris* (Brongniart) Brongniart, 1828 (Fern fronds)
*P. bourozii* Dalinval, 1960 ............E-125 (Text-fig. 79A)
*P. bucklandii* Brongniart, 1828 ............E-127 (Text-fig. 79E)
*P. intermedia* Dalinval, 1960 ............E-120 (Text-fig. 77B)
*P. lobulata* Dalinval, 1960 ............E-125 (Text-fig. 78C)
*P. plumosa* (Artis) Brongniart, 1828 ............E-129 (Text-figs 76A, 79C)
*P. unita* Brongniart, 1828 ............E-119, E-129, E-131 (Text-fig. 79D)
*P. volkmannii* Sauveur, 1848 ............E-120 (Text-fig. 79B)

*Polymorphopteris* Wagner, 1958 (Fern fronds)
*P. polymorpha* (Brongniart) Wagner, 1958 ............E-127
............(Pl. 22, fig. 3; Text-fig. 78A)

*Polypterocarpus* (Isolated ?pteridosperm seeds)
*P. anglicus* (Arber) Seward, 1917 ............H-10 (Text-fig. 86E)
*P. johnsonii* (Kidston) Crookall, 1976 ............H-5 (Text-fig. 86F)
*P. ornatus* (Arber) Kidston, 1914 ............H-10 (Text-fig. 86G)

*Potoniea* Halle, 1933 (Isolated pteridosperm pollen organs)
*P. carpentieri* (Kidston) Halle, 1933 ............H-34
............ (Pl. 29, fig. 11; Text-fig. 91A)

*Renaultia* Zeiller, 1883 (Fern fronds)
*R. chaerophylloides* (Brongniart) Zeiller, 1883 ....E-109 (Text-fig. 75D)
*R. crepinii* (Stur) Gothan, 1913 ............E-113 (Text-fig. 75B)
*R. footneri* (Marratt) Brousmiche, 1983 ............E-111
............(Pl. 25, fig. 2; Text-fig. 75F)
*R. gracilis* (Brongniart) Zeiller, 1883 ............E-116 (Text-fig. 75E)
*R. rotundifolia* (Andrä) Zeiller, 1899 ............E-108 (Text-figs 75A, 76B)
*R. schatzlarensis* (Stur) Danzé, 1956 ............E-99 (Text-fig. 75C)

*Reticulopteris* Gothan, 1941 (Pteridosperm leaves)
*R. muensteri* (Eichwald) Gothan, 1941 ............E-29 (Text-fig. 55D)

*Rhabdocarpus* Göppert and Berger, 1848 (Isolated pteridosperm seeds)
*R. renaultii* Kidston, 1914 ............H-27 (Text-fig. 89F)

# 220 British Coal Measures

*Samaropsis* Göppert, 1864 (Isolated seeds)
*S. bisecta* (Dawson) Crookall, 1976...............H-5 (Text-fig. 86D)
*S. crassa* (Lesquereux) Arber, 1914...............H-13 (Text-fig. 86C)
*S. emarginata* (Göppert and Berger) Kidston, 1911...............H-14
...............(Text-fig. 86A)
*S. gutbieri* (Geinitz) Kidston, 1914...H-14 (Pl. 29, fig. 5; Text-fig. 86B)

*Selaginellites* Zeiller, 1906 (Herbaceous lycopsid shoots)
*S. gutbieri* (Göppert) Kidston, 1911...............D-4 (Pl. 15, fig. 3)

*Sigillaria* Brongniart, 1822 (Arborescent lycopsid stems)
*S. boblayi* Brongniart, 1828...............B-42 (Pl 7, fig. 1; Text-fig. 36A)
*S. brardii* Brongniart, 1822...............B-29 (Text-fig. 37B)
*S. candollei* Brongniart, 1828...............B-31, B-40 (Text-fig. 35N)
*S. cordiformis* Kidston, 1911...............B-47 (Text-fig. 35F)
*S. cordigera* Zeiller, 1886...............B-55 (Text-fig. 35H)
*S. davreuxii* Brongniart, 1828...............B-37 (Text-fig. 35D)
*S. distans* Sauveur, 1848...............B-53 (Text-fig. 35K)
*S. elegans* (Sternberg) Brongniart, 1828...............B-28 (Text-fig. 37E)
*S. elongata* Brongniart, 1824...............B-32 (Text-fig. 36H)
*S. ichthyolepis* (Presl) Corda, 1845...............B-29 (Text-fig. 37D)
*S. kidstonii* Crookall, 1925...............B-46 (Text-fig. 36L)
*S. laevigata* Brongniart, 1828...............B-53 (Text-fig. 35I)
*S. latibasa* Crookall, 1966...............B-43 (Text-fig. 36G)
*S. lorwayana* Dawson, 1873...............B-54 (Pl. 8, fig. 2; Text-fig. 37A)
*S. macmurtriei* Kidston, 1885...............B-8 (Text-fig. 37C)
*S. mamillaris* Brongniart, 1824...............B-28 (Text-fig. 35C)
*S. micaudii* Zeiller, 1886...............B-38 (Text-fig. 35E)
*S. nortonensis* Crookall, 1925...............B-47 (Pl. 8, fig. 1; Text-fig. 36B)
*S. nudicaulis* Boulay, 1876...............B-50 (Text-fig. 35J)
*S. ovata* Sauveur, 1848...............B-55 (Pl. 7, fig. 2; Text-fig. 36I)
*S. polyploca* Boulay, 1876...............B-35 (Text-fig. 35B)
*S. principis* Weiss, 1882...............B-39 (Text-fig. 36E)
*S. reniformis* Brongniart, 1824...............B-38 (Text-fig. 36D)
*S. reticulata* Lesquereux, 1866...............B-57 (Text-fig. 37F)
*S. rugosa* Brongniart, 1828...............B-32 (Pl. 7, fig. 5; Text-fig. 35M)
*S. sauveurii* Zeiller, 1886...............B-46 (Text-fig. 36K)
*S. schlotheimiana* Brongniart, 1836...............B-52 (Text-fig. 35L)
*S. scutellata* Brongniart, 1822...............B-40 (Pl. 7, fig. 3; Text-fig. 36F)
*S. scutiformis* Zalessky, 1904...............B-50 (Text-fig. 35G)
*S. sol* Kidston, 1897...............B-47 (Text-fig. 36C)
*S. tessellata* Brongniart, 1828...............B-44 (Pl. 7, fig. 4; Text-fig. 36M)
*S. transversalis* Brongniart, 1828...............B-44 (Text-fig. 36J)
*S. youngiana* Kidston, 1894...............B-52 (Text-fig. 35A)

# Taxonomic Index

*Sigillariostrobus* (Schimper) Feistmantel, 1876
(Arborescent lycopsid cones)
   *S. rhombibracteatus* Kidston, 1897 ............................G-4 (Pl. 28, fig. 2)
   *S.* sp. ................................................................................................F-3

*Sphenophyllum* Brongniart, 1822 (Sphenophyll leaves)
   *S. cuneifolium* (Sternberg) Zeiller, 1880 ...............................C-17, C-20
   ............................................................................(Pl. 13, fig. 1; Text-fig. 46E)
   *S. emarginatum* Brongniart, 1828...........C-15, C-16, C-20 (Text-fig. 45)
   *S. majus* (Bronn) Bronn, 1835 .................................C-17 (Text-fig. 46B)
   *S. myriophyllum* Crépin, 1880.................................C-19 (Text-fig. 46C)
   *S. oblongifolium* (Germar and Kaulfuss) Unger, 1850......................C-12
   .............................................................................................(Text-fig. 46D)
   *S. trichomatosum* Stur, 1887........................C-15, C-18 (Text-fig. 46A)

*Sphenopteris* (Brongniart) Sternberg, 1825 (Fern or pteridosperm leaves)
   *S. andraeana* von Roehl, 1868...............................E-65 (Text-fig. 65A)
   *S. macilenta* Lindley and Hutton, 1831...................E-63 (Text-fig. 65B)
   *S. rutaefolia* Gutbier, 1835....................................E-110 (Text-fig. 73G)
   *S. schwerinii* (Stur) Zeiller, 1889............................E-110 (Text-fig. 73I)
   *S. selbyensis* Kidston, 1923 ...................................E-116 (Text-fig. 73H)
   *S. sewardii* Kidston, 1923 .....................................E-102 (Text-fig. 73D)
   *S. souichii* Zeiller, 1888...........................................E-98 (Text-fig. 73J)

*Sphyropteris* Kidston, 1889 (Fern fronds)
   *S. obliqua* (Marratt) Kidston, 1889 .......................E-113 (Text-fig. 74H)

*Stigmaria* Brongniart, 1822 (Arborescent lycopsid rooting organs)
   *S. ficoides* (Sternberg) Brongniart, 1822 ............................................B-61
   ...............................................................................(Pl. 9, fig. 2; Text-fig. 39C)
   *S. stellata* Göppert, 1841........................................B-61 (Text-fig. 39D)
   *S.* sp. ...........................................................................D-2 (Pl. 9, fig. 1)

*Sturia* Němejc, 1934 (Fern fronds)
   *S. amoena* (Stur) Němejc, 1934..............................E-115 (Text-fig. 73F)

*Sublepidophloios* Sterzel, 1907 (Lycopsid stems)
   *S. ventricosus* Hopping, 1956..................................B-24 (Text-fig. 34E)

*Syringodendron* Sternberg, 1820 (decorticated lycopsid stem)
   *S.* sp. ........................................................................B-1 (Text-fig. 28A)

*Trigonocarpus* Brongniart, 1828 (Isolated pteridosperm seeds)
   *T. candollianus* Unger, 1870 ..................................H-25 (Text-fig. 89D)

## 222 British Coal Measures

*T. noeggerathii* (Sternberg) Brongniart, 1828 .........H-24 (Text-fig. 89B)
*T. oliveriformis* Crookall, 1976 ....................H-25 (Text-fig. 89A)
*T. parkinsonii* Brongniart, 1828 .................................H-24
............................................ (Pl. 29, figs 3, 4, 6; Text-fig. 89C)

*Ulodendron* Lindley and Hutton, 1831 (Lycopsid stems and leafy shoots)
  *U. landsburgii* (Kidston) Thomas, 1968 ...................D-7 (Text-fig. 47B)
  *U. majus* Lindley and Hutton, 1831 ..............D-7 (Pl. 10; Text-fig. 47A)

*Urnatopteris* Kidston, 1884 (Fern fronds)
  *U. herbaceae* (Boulay) Kidston, 1923 ...............E-102 (Text-fig. 74A)

*Walchia* Sternberg, 1825 (Conifer twigs)
  *W.* sp. .....................................................D-3 (Pl. 15, figs 1,2)

*Whittleseya* Newberry, 1873 (Isolated pteridosperm pollen organs)
  *W.* sp. ....................................H-35 (Pl. 29, fig. 1; Text-fig. 89E)

*Zeilleria* Kidston, 1884 (Fern fronds)
  *Z. delicatula* (Sternberg) Kidston, 1884 ..............E-100 (Text-fig. 74C)
  *Z. frenzlii* (Stur) Gothan, 1913 ......................E-103 (Text-fig. 74D)
  *Z. hymenophylloides* Kidston, 1924 ..................E-115 (Text-fig. 74B)

Miscellaneous
  Halonial scars (Lycopsid stems) ......................B-62 (Text-fig. 40A)
  Ulodendroid scars (Lycopsid stems) .........................B-62 (Pl. 10)